PEARSON

GEOGRAPHY

NEW SOUTH WALES

AUTHORS
Andre Chadzynski
Alon Kaiser
Andrew Peters

7+8

STAGE 4 • ACTIVITY BOOK

Pearson Australia
(a division of Pearson Australia Group Pty Ltd)
707 Collins Street, Melbourne, Victoria 3008
PO Box 23360, Melbourne, Victoria 8012
www.pearson.com.au

First published 2013 by Pearson Australia
2020 2019 2018 2017
10 9 8 7 6 5 4 3

Content & Learning Specialist: Marita Tripp
Project Manager: Michelle Thomas
Production Manager: Anji Bignell
Editor: Lucy Heaver
Designer: Anne Donald
Rights & Permissions Editor: Lisa Woodland
Senior Publishing Services Analyst: Rob Curulli
Typesetter: Aptara
Illustrators: Guy Holt and Bruce Rankin
Printed in Australia by the SOS Print + Media Group

ISBN: 978 1 4886 1392 0

Pearson Australia Group Pty Ltd ABN 40 004 245 943

Contents

Contents continued

How to use this book

PEARSON geography NSW Stage 4 Activity Book

PEARSON geography NSW Stage 4 Activity Book is an integral part of the Australian Curriculum series PEARSON geography NSW. It caters for a variety of learning styles and will reinforce, extend and enrich learning initiated through the student book.

PEARSON geography NSW Stage 4 Activity Book is designed for independent use by students and is suitable for in-class use or as a complete homework program.

Activity Book	Corresponding Student Book chapter
Chapter 1	Chapter 1: Geography's tools and skills Chapter 18: Geoskills
Chapter 2	Chapter 2: Landscapes and landforms Chapter 4: Coastal landforms Chapter 15: Alpine landforms Chapter 16: Riverine landforms Chapter 17: Desert landforms
Chapter 3	Chapter 3: Landscape change, management and protection
Chapter 4	Chapter 5: Geomorphic hazards
Chapter 5	Chapter 6: Place and liveability

Activity Book	Corresponding Student Book chapter
Chapter 6	Chapter 7: Enhancing liveability
Chapter 7	Chapter 8: Water: A renewable natural resource
Chapter 8	Chapter 9: Australia's water resources
Chapter 9	Chapter 10: Water scarcity and management
Chapter 10	Chapter 11: Atmospheric and hydrologic hazards
Chapter 11	Chapter 12: Personal connections
Chapter 12	Chapter 13: Technology connecting people and places
Chapter 13	Chapter 14: Production, consumption and trade

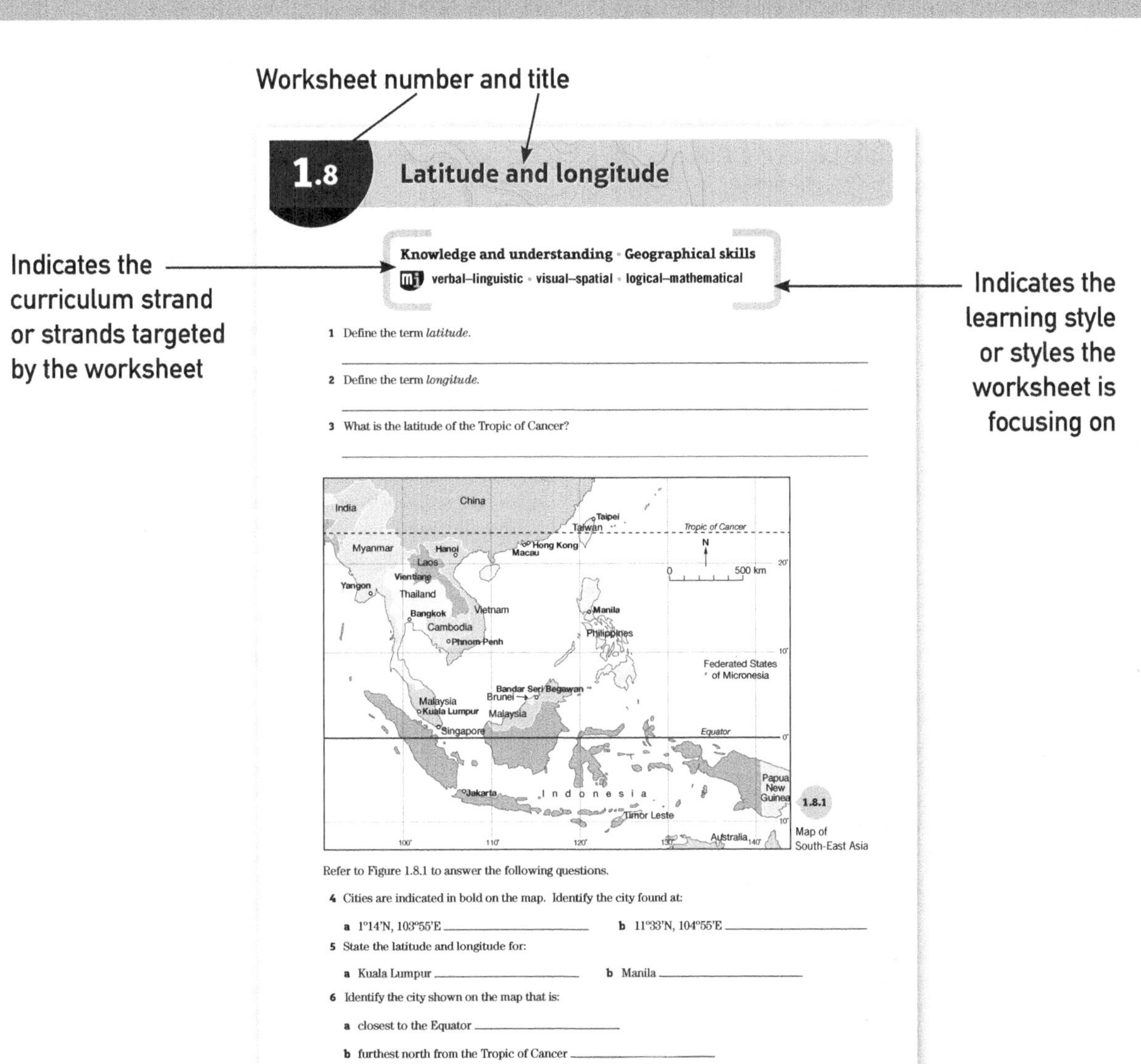

1.8 Latitude and longitude

Knowledge and understanding • Geographical skills
verbal–linguistic • visual–spatial • logical–mathematical

1 Define the term *latitude*.

2 Define the term *longitude*.

3 What is the latitude of the Tropic of Cancer?

1.8.1 Map of South-East Asia

Refer to Figure 1.8.1 to answer the following questions.

4 Cities are indicated in bold on the map. Identify the city found at:

a 1°14'N, 103°55'E ______ b 11°33'N, 104°55'E ______

5 State the latitude and longitude for:

a Kuala Lumpur ______ b Manila ______

6 Identify the city shown on the map that is:

a closest to the Equator ______

b furthest north from the Tropic of Cancer ______

Acknowledgements

We thank the following for their contributions to our text book:

Cover image: Walter Bibikow/Getty Images

We would like to thank the following for permission to reproduce copyright material. The following abbreviations are used in this list: t = top, b = bottom, l = left, r = right, c = centre.

AAP: AP/Kyodo News, p. 48l.
ABC: ABC Rural: 'White gold rush - Milk formula exports to China', by Sarina Locke, 4 June 2013, p.140.
ABS: © Commonwealth of Australia, Australian Bureau of Statistics - 3101.0, Australian Demographic Statistics, Jun 2012, p. 66; 2071.0 , Reflecting a Nation: Stories from the 2011 Census, 2012–2013, p. 68, p. 129.
Alamy Ltd: Robert Bush, p. 20c; dbimages, p. 121t; Sindre Ellingsen, p. 121b; Gallo Images, p. 33tr; GC Stock, p. 30br; Suzanne Long, p. 132; McClatchy-Tribune Information Services, p. 131; one-image photography, p. 28; Frantisek Staud, p. 33tl; Doug Steley, p. 1l; Jack Sullivan, p. 42t; World Travel Collection, p. 30bl.
Australian Heritage Commission: *Nourishing Terrains, Australian Aboriginal Views of Landscape and Wilderness*, Deborah Bird Rose, Australian Heritage Commission, 1996, ISBN 0 642 23561 9, p. 15t; p. 15b; p.16
Bio Diversity BC: http://www.biodiversitybc.org/, p. 145.
Hugh Brown: p. 13.
City of Sydney: p. 128.
Corbis: Susan Dykstra/Design Pics, p. 19b; Imaginechina, p. 135t; Atlantide Phototravel, p. 118t.
Copyright Clearance Centre Inc: *Pacific Standard Magazine*, 'The Music Festival Bubble', by Sarah Sloat, 14 June 2013, p. 132.
Conservation International: p. 144.
Department of Natural Resources and Mines: Queensland Government, p. 94.
Dreamstime: Michele Alfieri, p. 36tr; Ashestosky, p. 139; Dan Breckwoldt, p. 25bc; Terry Carr, p. 34bl; Chunli Li, p. 25t; Florentino David, p. 14t; Alen Dobric, p. 14br; Aleksandra Durdyn, p. 33br; Echomech, p. 72; Darla Hallmark, p. 36b; Howardliuphoto, p. 20b; Kurhan, p. 14cl; Lisa Cora Reed, p. 14cr; Robhainer, p. 14bl; Srlee2, p. 49t; Rechitan Sorin, p. 35br; Nickolay Stanev, p. 35tl; Stefanschurr226, p. 36tl; Tupungato, p. 71; Vvuls, p. 22; Wangkun Jia, p. 31tr; Keith Wheatley, p. 33bl.
Environmental Working Group: www.ewg.org. Reprinted with permission, p. 147.
Epson Australia: p. 143.
European Commission's Joint Research Centre and the World Bank: p. 135b.
Getty Images: Dr Juerg Alean, p. 25b; flashfilm, p. 118b; Carlyn Iverson, p. 25tc; Mark Ralston/AFP, p. 47b.
GRID-Arendal: Philippe Rekacewicz, p. 8.
Hardie Grant/UBD: Map section reproduced with permission of UBD-Gregory's. Copyright Hardie Grant Media Pty Ltd DD 06/16, p. 6r.
Josh Hill: p. 19c.
IRIN: IRINnews.org, p. 101.
Keyport Online: p. 120.
MagpieShooter: p. 137.
Many Possibilities: Stephen Song, manypossibilities.net, p. 138.
NASA Images: SA/courtesy of nasaimages.org, p. 31tl;
National Park Service: p. 49b; p. 117.
NSW Department of Lands: © LPI - Department of Finance and Services [year], Panorama Avenue, Bathurst 2795, p. 6l.
OECD Publishing: Net agricultural production for world and economic groups, Net production index 2010-19, OECD-FAO Agricultural Outlook http://www.oecd.org/site/oecd-faoagriculturaloutlook/netproductionindex2010-19.htm, p. 142b.
Oz Aerial: p. 43l.
Parks Australia: Department of the Environment and Water Resources, Director of National Parks, p. 41.
Pearson Asset Library: iofoto, p. 85.
Pearson Australia: Siân Bradfield, p. 19t.
The Phnom Penh Post: 'Hun Sen announces new airport', by Rann Reuy, p. 29 Oct 2012, p. 125; 'Cambodian Resorts on the Increase', by Rupert Winchester, p. 4 Oct 2012, p. 125.
Chris Renk Photography: www.chrisrenk.com, p. 49c.
Project for Public Spaces Inc: Reprinted with permission © 2014 All Rights Reserved, p. 120.
Radio Free Asia: © 1998-2013, RFA. Used with the permission of Radio Free Asia, 2025 M St. NW, p. Suite 300, Washington DC 20036. http://www.rfa.org., p. 127.
The Register: 'Australian volcano starts to blow', by Simon Sharwood, 24th October 2012, p. 50.
Rural Advancement Foundation International – USA: p. 148.
Shutterstock: Dustie, p. 110b; dymax, p. 78; edella, p. 5; feiyuezhangjie, p. 1r; jiewphoto, p. 39tr; Juancat, p. 30tr; Alexander Kazantsev, p. 39bl; mamahoohooba, p. 39br; Palo_ok, p. 39tl; Scott Prokop, p. 35tr; Portadown, p. 69; Mariusz Szczygiel, p. 4r; Vinicius Tupinamba, p. 4b; Peter Wollinga, p. 42b; A.S. Zain, p. 8l; 1000 Words, p. 73.
Skyepics.com.au: p. 42r.
The (US) National Science Foundation: National Center for Science and Engineering Statistics, p. 142t.
The Telegraph Media group Ltd: Adapted from *The Telegraph*, 'Scores killed by eruption from Indonesian volcano', p. 5 November 2010, p. 51; 'Australia's surfers mourn disappearing east coast beaches as currents sweep sand out to sea' by Jonathan Pearlman, *The Telegraph*, p. 25 Feb 2012, p. 43.
University of British Columbia: p. 123;
University of Melbourne: Dr Wayne Atkinson for Yorta Yorta Native Title Claim, 1997, p. 96.
Western Australian Planning Commission: Artwork adapted from Western Australian Planning Commission Department of Planning Western Australia. Figures 3 and 4 landscaping obscuring/enabling surveillance, p. 58; Figures 5/6 poor/good urban structure, p. 58.

Every effort has been made to trace and acknowledge copyright. However, should any infringement have occurred, the publishers tender their apologies and invite copyright owners to contact them.

Disclaimer/s

Some of the images used in *Pearson Geography NSW Stage 4 Activity Book* might have associations with deceased Indigenous Australians. Please be aware that these images might cause sadness or distress in Aboriginal or Torres Strait Islander communities.

1.1 What is geography?

Knowledge and understanding • Geographical skills

mi **verbal–linguistic • visual–spatial**

1 Give two reasons why it is important to study geography.

____

____

2 Use words from the box below to complete each of the statements about geography.

features	people	processes	world	places	events	environments	characteristics

a Geography is the study of the ____________________ that make up our world.

b Geography is concerned with the ____________________ that shape the earth's surface.

c Geography studies the way people interact with their ____________________.

d Geography seeks to explain the character of ____________________ and the distribution of ____________________, ____________________ and ____________________ on or near the earth's surface.

e Geography helps us to better understand the ____________________ in which we live.

Study the photographs in Figures 1.1.1 and 1.1.2 and then answer the questions that follow.

1.1.2 Pollution from stormwater released into the Yarra River, Melbourne, Victoria

1.1.1 A giant open-cut coal mine in the Bowen Basin, Queensland

3 The questions geographers ask are called geographical questions. Geographical questions include things like 'How is it changing?' and 'What might it be like in the future?'

a Write two geographical questions for Figure 1.1.1.

____

____

____

____

b Write two geographical questions for Figure 1.1.2.

____

____

____

____

4 **a** In the table below list four physical and four human characteristics about the suburb or neighbourhood where you live.

Name of suburb/neighbourhood: ______________________________

Physical characteristics	Human characteristics
•	•
•	•
•	•
•	•

b Explain whether this a geographical question: 'If mining continues in Kakadu National Park, what might happen to the park in the future?'

____

____

1.2 Exploring the world

Knowledge and understanding • Geographical skills

 verbal–linguistic • visual–spatial

1 Use words from the box below to correctly label the continents, oceans and lines of latitude on the world map.

Africa	Europe	North America	South America	Atlantic Ocean
Asia	Antarctica	Southern Ocean	Pacific Ocean	Arctic Ocean
Australia	Indian Ocean	Tropic of Cancer	Tropic of Capricorn	Equator

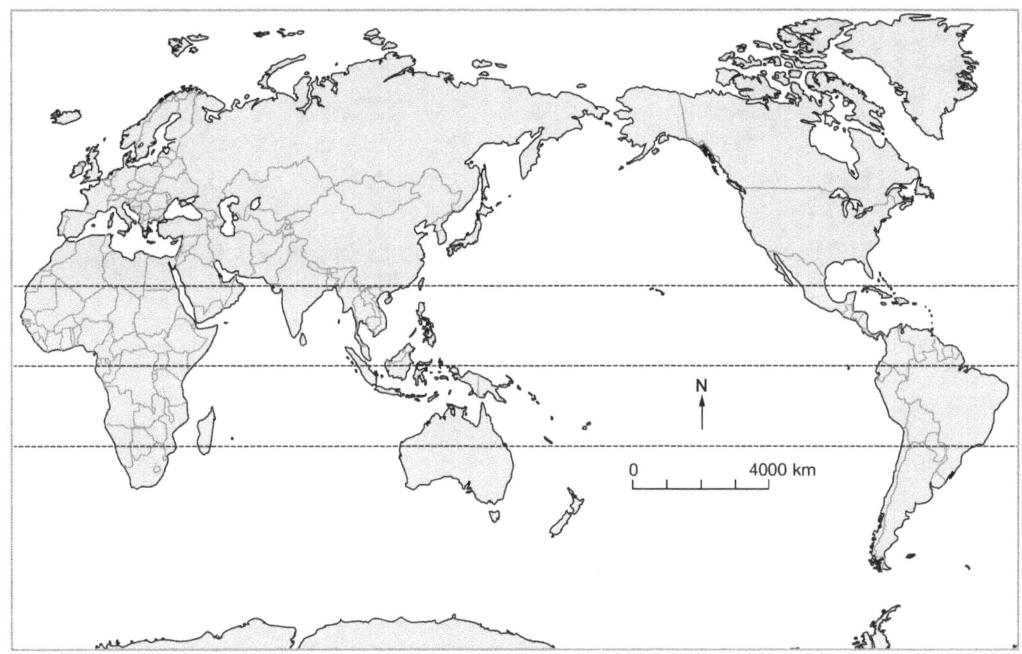

2 Name the continent that matches each of the descriptions below.

a Made up of one large island and many smaller islands scattered throughout the south western part of the Pacific Ocean. ____________________

b A large continent in the northern hemisphere, between the Pacific Ocean and the Atlantic Ocean.

____

c A large continent that spans both the northern and southern hemispheres. On one side is the Indian Ocean and on the other, the Atlantic Ocean. ____________________

d A large continent, entirely in the southern hemisphere, and separated from all other continents and surrounded by the Southern Ocean. ____________________

e A small continent in the northern hemisphere, which is separated from another larger continent by a mountain range. ____________________

f A large continent found mostly in the northern hemisphere that is connected to both the Pacific Ocean and the Indian Ocean. ____________________

g A large continent in the southern hemisphere between the Pacific Ocean and the Atlantic Ocean.

____

1.3 Geoskills: Analysing photographs

Geographical skills

 visual–spatial

1 Study the photographs in Figures 1.3.1 to 1.3.3. For each photograph, state whether the perspective is: *ground level*, *aerial* or *oblique*.

1.3.1 Buildings damaged by the 2004 tsunami in Aceh, Indonesia

Type of photograph: ____________________

1.3.2 City centre, Otmuchow, Poland

Type of photograph: ____________________

1.3.3 View of London with Houses of Parliament in the foreground

Type of photograph: ____________________

2 Analyse the photograph in Figure 1.3.1 and complete the table below.

Type of environment (natural, managed or constructed)	
Main features shown	
Evidence of location and time (from the photograph itself, the caption or the source information)	
Features out of place or unexpected	

Practise your photo sketching by drawing a sketch from the photograph in Figure 1.3.4.

1.3.4

Kata Tjuta, formerly known as the Olgas, in the Northern Territory

1.4 Types of maps

Knowledge and understanding

 **verbal–linguistic • visual–spatial**

Classifying maps

1 a Label the following maps as street or topographic maps.

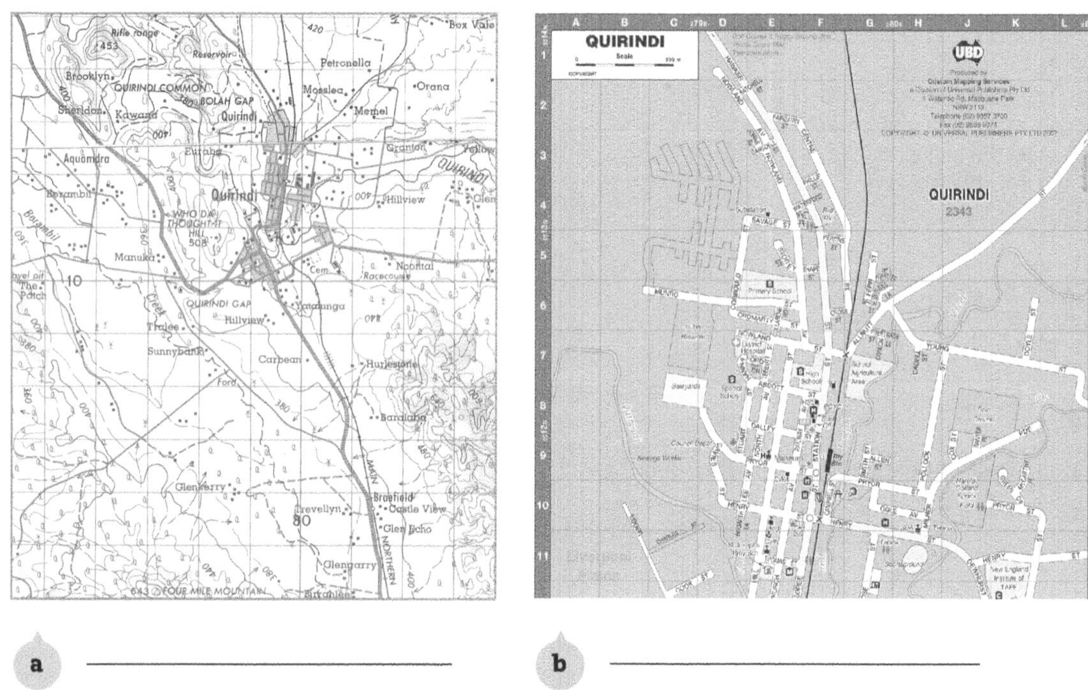

a ______________________ b ______________________

b In the Venn diagram, list the differences and similarities between the maps.

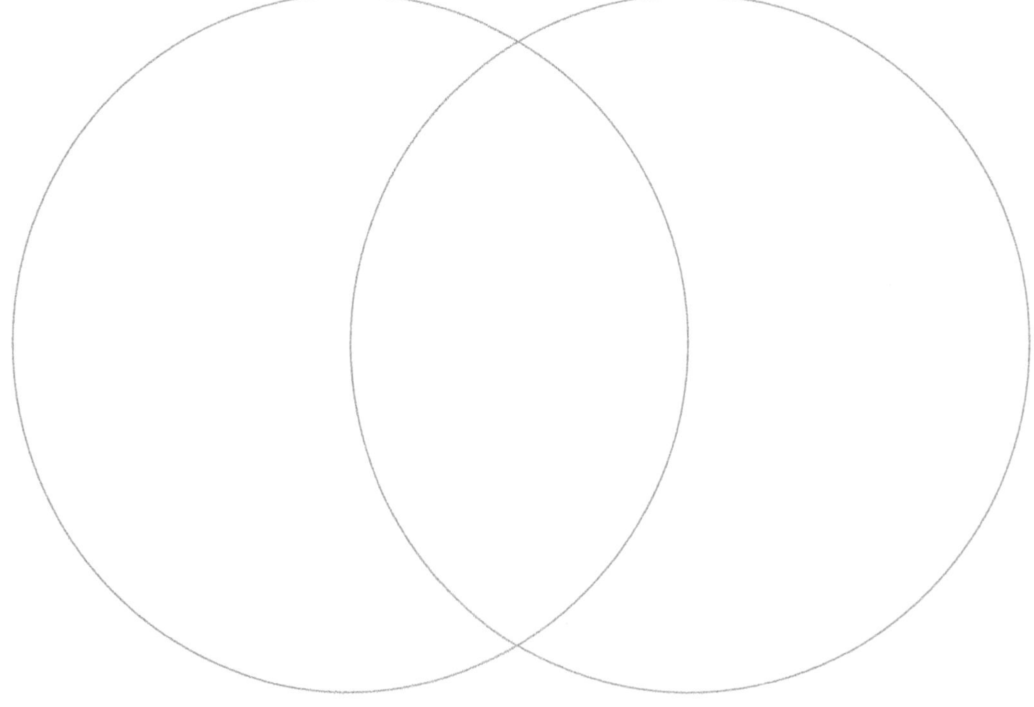

2 List two key features of each type of map in the table below.

Map type	Key features
Topographic	
Thematic	
Weather	
Street	

3 For each of the following situations, identify the type of map needed and provide a reason for your decision.

a Tammy needs to find her way to her friend's place on the other side of the city.

b Pratik, Karl and Jo are hiking from Jo's farm to a nearby creek.

c Justin has been asked by his boss to identify an area of high population growth.

d The local council is investigating where park and playing fields are located.

1.5 Elements of a map

Knowledge and understanding

verbal–linguistic • visual–spatial

1 Use the map in Figure 1.5.1 to complete the questions that follow.

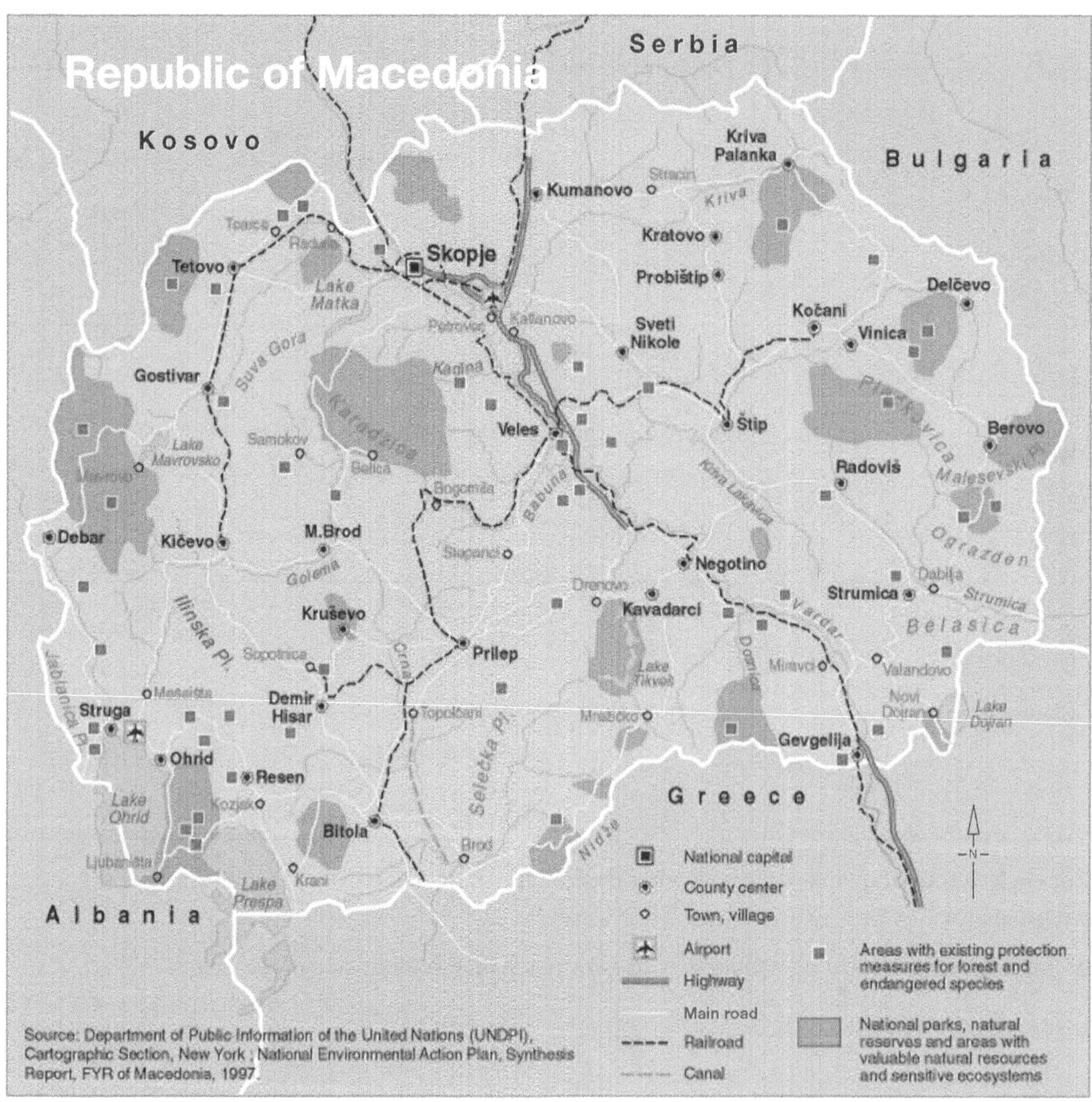

1.5.1 Map of the Former Yugoslav Republic of Macedonia

a Identify which aspect of BOLTSS (border, orientation, legend, title, source, scale) is missing from the map.

____

b List four main types of transport used in the region.

i ______________ **ii** ______________

iii ______________ **iv** ______________

c Discuss what you think is the main purpose of this map.

____

____

d Rob and Jess are Australian tourists in the Republic of Macedonia. They are having trouble reading their map and would like you to help them. They want to travel from Gevgelija to Kavadarci to visit relatives, then go to the capital and from there travel to a spa in Struga. They show you the map and ask what transport options they have for their journey.

Use the map to complete the flow chart below about their transport options.

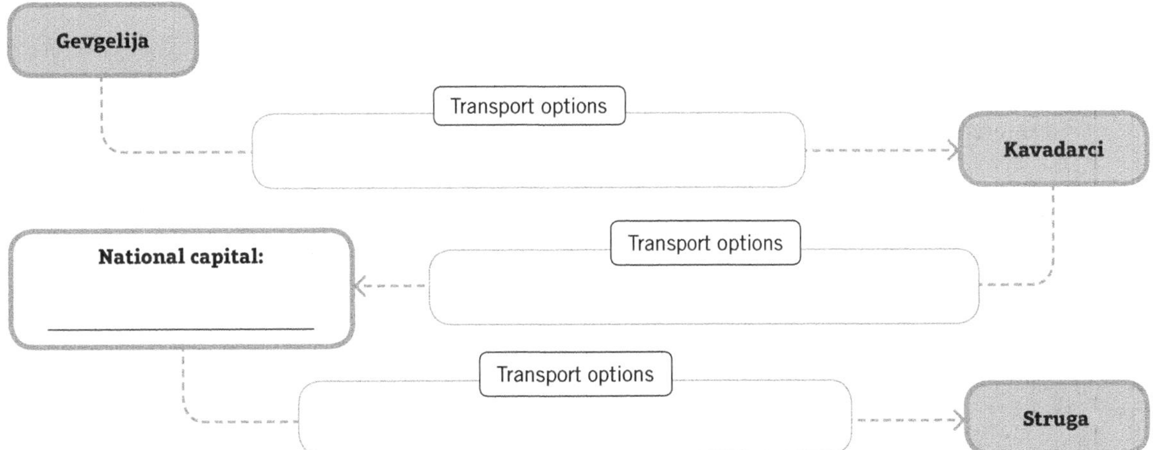

2 What might each of the following map symbols represent? Write your answer in the space provided.

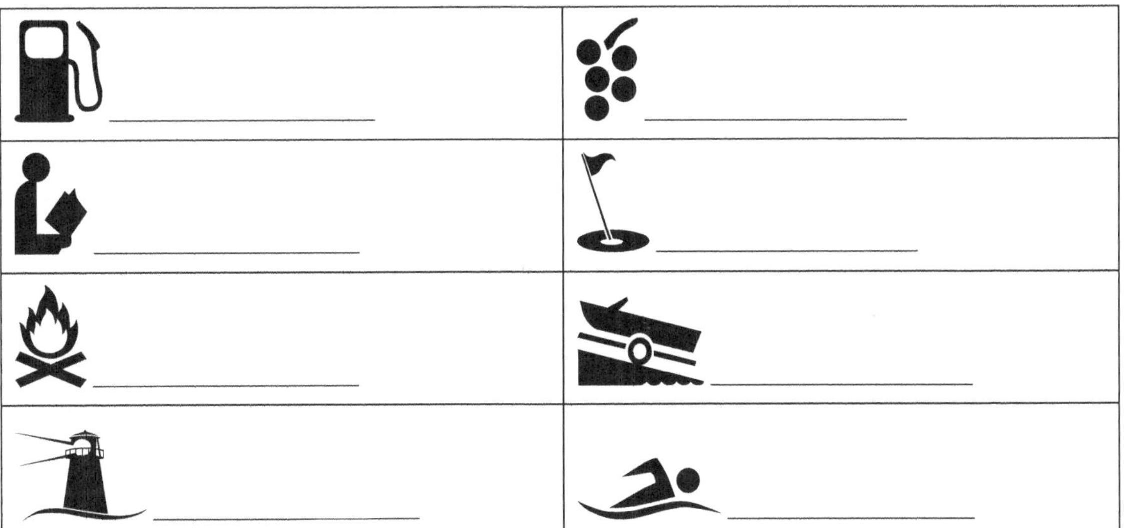

3 In the space below, draw a map of a sports field that is 100 metres long by 50 metres wide. Decide on the best scale to use and write it on your map. Make sure your map includes BOLTSS.

1.6 Grid and area references

Knowledge and understanding • Geographical skills

mi verbal–linguistic • visual–spatial • logical–mathematical

1 Define grid and area references. Explain why each may be used.

Grid reference: ______________________________

____

____

Area reference: ______________________________

____

____

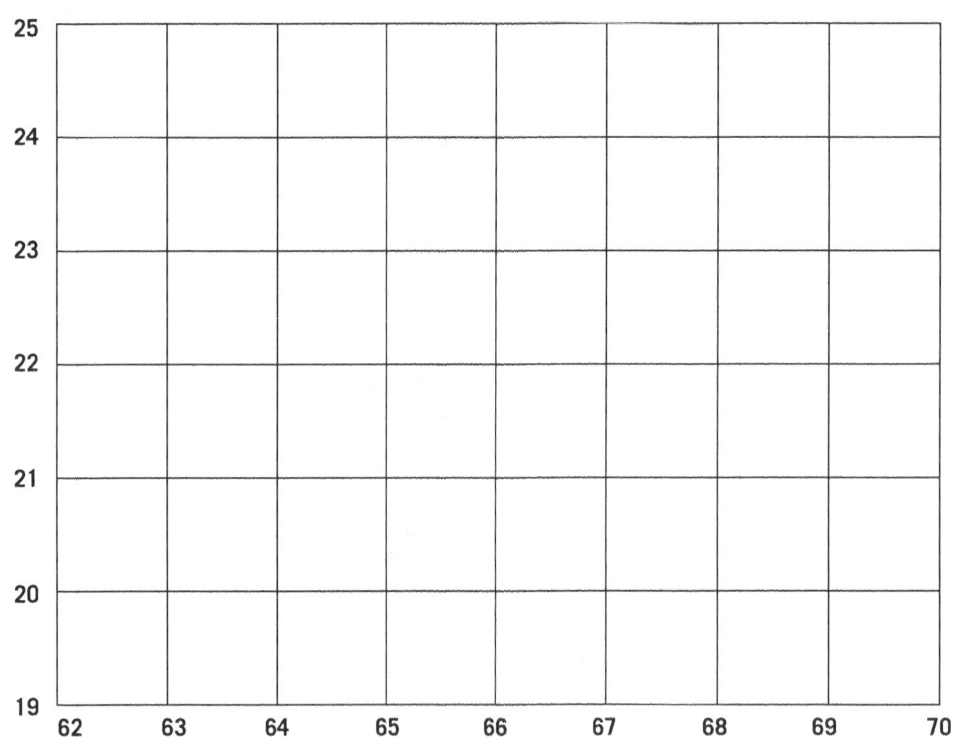

2 On the grid, shade the following area references in the colour indicated:

a AR 6522—red **b** AR 6424—blue **c** AR 6520—yellow **d** AR 6923—green

3 Choose a symbol for each of the following terms and record it in the box provided. Then, on the grid, place the appropriate symbol at the grid reference indicated.

a Mine—GR 625225

b Airport—GR 648238

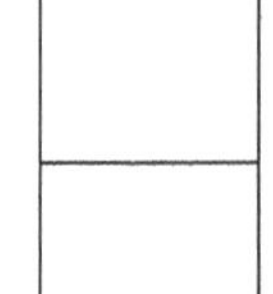

c Post office—GR 690200

d Hospital—GR 661191

1.7 Geoskills: Topographic maps

Geographical skills

 **logical–mathematical • visual–spatial**

1 Describe the purpose of a topographic map.

____

2 Explain what a contour map shows.

____

Use the simple topographic map below to answer the following questions.

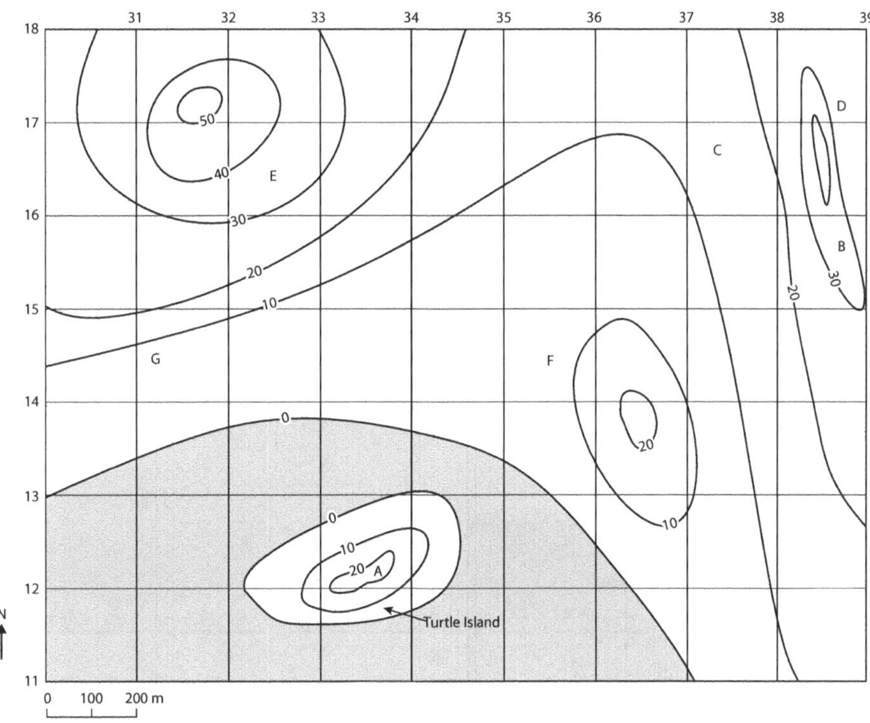

3 Identify the contour interval of this map. ____

4 Describe the relief between points A and B. ____

____

5 Identify the approximate height of point C. ____

6 Identify the area reference of the map where the slope is steepest. ____

7 If you were sitting at point A, would you be able to see another person sitting at point D? __________

Explain your reasoning. ____

____

8 Identify the aspect of the slope at point E. ____

9 Identify the direction of F from each of the following points.

a A: __________ **b** C : __________ **c** E: __________ **d** G: __________

1.8 Latitude and longitude

Knowledge and understanding • Geographical skills

mi verbal–linguistic • visual–spatial • logical–mathematical

1 Define the term *latitude*.

____

2 Define the term *longitude*.

____

3 What is the latitude of the Tropic of Cancer?

____

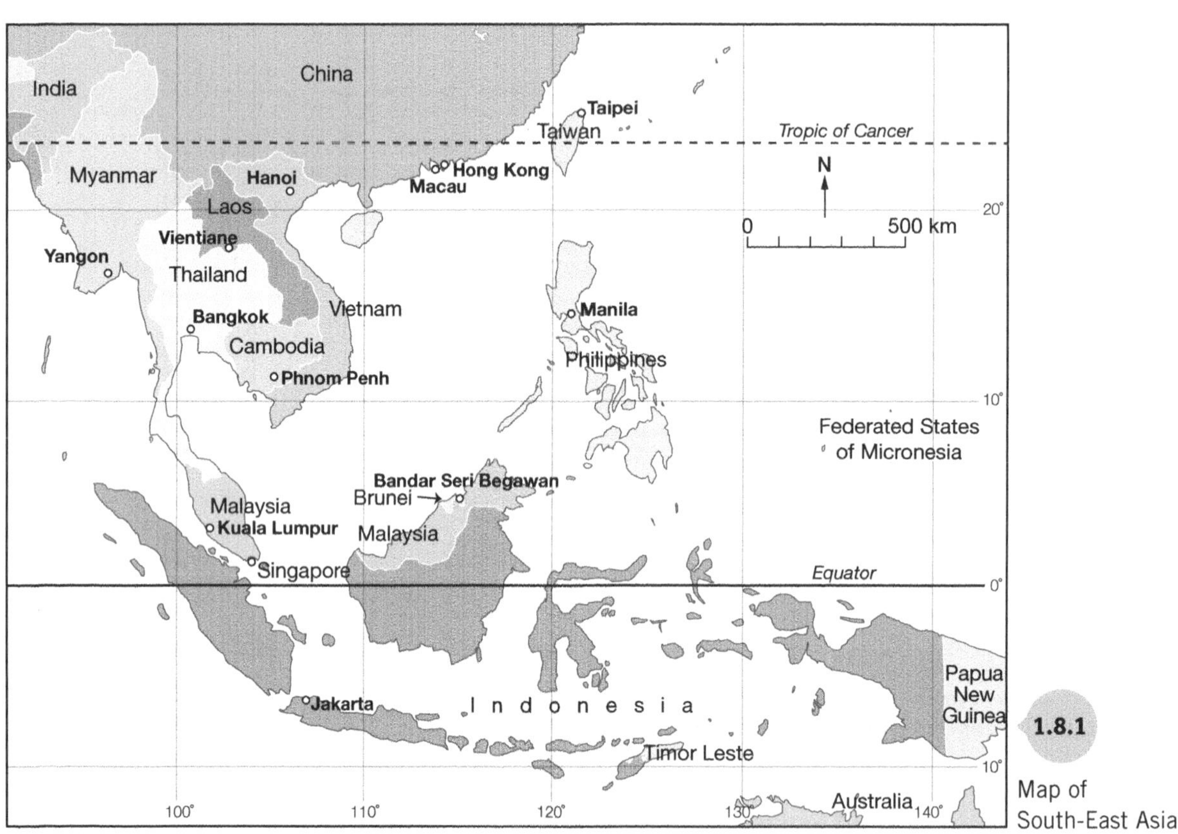

1.8.1 Map of South-East Asia

Refer to Figure 1.8.1 to answer the following questions.

4 Cities are indicated in bold on the map. Identify the city found at:

a 1°14'N, 103°55'E ______________ **b** 11°33'N, 104°55'E ______________

5 State the latitude and longitude for:

a Kuala Lumpur ______________ **b** Manila ______________

6 Identify the city shown on the map that is:

a closest to the Equator ______________

b furthest north from the Tropic of Cancer ______________

2.1 Landscapes

Knowledge and understanding • Geographical skills

 **visual–spatial • verbal–linguistic**

'Landscape' is the term used to describe the visible features of an area. These features are called 'elements' and include natural, living, human and changeable elements.

2.1.1 Ord River, Kununurra, Kimberley region, Western Australia

1 Describe types of elements in the landscape in Figure 2.1.1.

a ______________________________

b ______________________________

c ______________________________

d ______________________________

2 Landscapes with many natural features often have a cultural overlay. They can be a combination of natural, managed and constructed elements. Discuss whether the landscape in Figure 2.1.1 is natural, managed or constructed, or a combination.

____

____

3 Describe the sense of place in Figure 2.1.1 by explaining why humans might enjoy being at this location.

____

____

____

2.2 Valuing landscapes

Knowledge and understanding • Geographical skills

 **visual–spatial • verbal–linguistic • intrapersonal**

2.2.1

Grand Canyon, USA

Landscapes may be valued in many ways. The Grand Canyon in the USA is a famous landscape, managed by the Grand Canyon National Park and two indigenous tribal groups. The Canyon is 446 km long, up to 29 km wide and in places as deep as 1.8 km. The Canyon was formed as the Colorado River cut through the rock, revealing spectacular rock formations and nearly two billion years of the earth's geological history.

1 For each of the four people below, expand on their opening sentence about how they value the Grand Canyon.

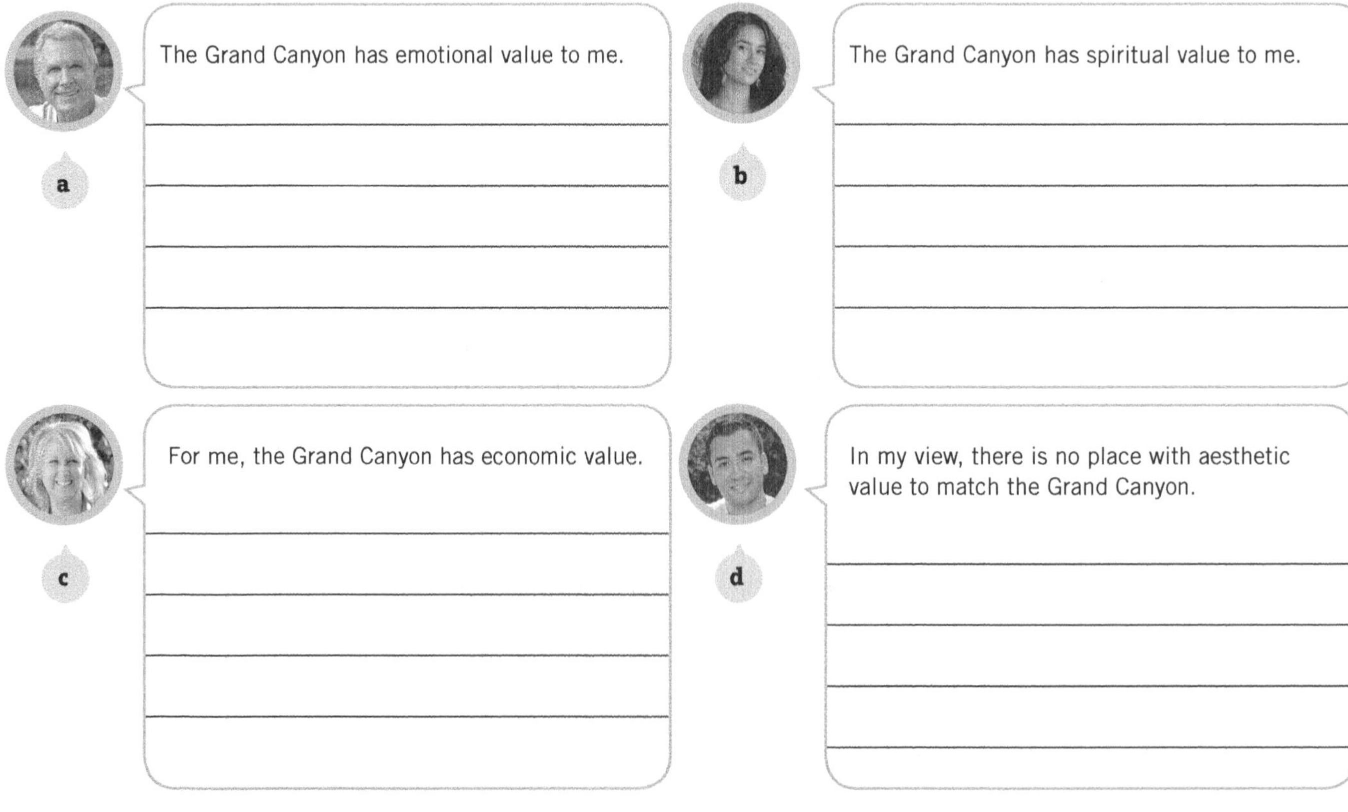

2.3 Indigenous explanations of landscapes

Knowledge and understanding • Geographical skills

mi visual–spatial • verbal–linguistic

1 Read the extract below then answer the questions that follow.

Sacred geography

The Australian continent is criss-crossed with the tracks of the Dreamings: walking, slithering, crawling, flying, chasing, hunting, weeping, dying, giving birth. Performing rituals, distributing the plants, making the landforms and water, establishing things in their own places, making the relationships between one place and another. Leaving parts or essences of themselves, looking back in sorrow; and still travelling, changing languages, changing songs, changing skin. They were changing shape from animal to human and back to animal and human again, becoming ancestral to particular animals and particular humans. Through their creative actions they demarcated a whole world of difference and a whole world of relationships which cross-cut difference.

Kakawuli (bush yam) *come up from Dreaming. No matter what come up, they come out of Dreaming. All tucker come out from Dreaming. Fish, turtle, all from Dreaming. Crocodile, anything, all come from Dreaming. Kangaroo, makaliwan* (wallaby), *all birds, all from Dreaming.*

Big Mick Kankinang

Source: D.B. Rose, 1996, *Nourishing Terrains: Australian Aboriginal Views of Landscape and Wilderness*

a In your own words, write the meaning of 'the Dreaming'.

____

____

b How do Aboriginal people explain the origins of the landscape?

____

____

____

The belief of Aboriginal people that the landscape is the result of processes, rather than just having always existed, is similar to the modern understanding. For them, the processes that formed the landscape have a spiritual explanation. It is not surprising, therefore, that they have a spiritual connection to the land.

2 Read the extract below then fill in the table that follows.

Kuku-Yalanji people of the rainforest of North Queensland tell of the origins of ceremony:

Long ago, when all the land was flat, Kurriyala came from the west. His first stopping place was Narabullgan, Mount Mulligan, which he formed out of his droppings. On top of this granite mountain he made a huge lake, then a deep gorge as he crawled away. He moved on to another big mountain which he made in the form of a snake. He called it 'Naradunga', now often known as Mount Mulgrave. Then he moved further north until he came to Fairview where he made a huge mountain out of lime and a big lagoon. After this he went northwards to his own people to a place called Bushy Creek. He saw his people dancing but they were not painted properly or dressed correctly so he came out from hiding and showed his people how to paint their bodies and how to dance.

Source: D.B. Rose, 1996, *Nourishing Terrains: Australian Aboriginal Views of Landscape and Wilderness*

Explain each landscape feature according to the Kuku-Yalanji people.

Landscape feature	Explanation by the Kuku-Yalanji people
Mount Mulligan	
Huge lake	
Deep gorge	
Mount Mulgrave	
Fairview	

3 Read the extract below then answer the questions that follow.

Many species of animals that are endangered are in that state because of habitat loss, and cessation of Aboriginal burning is one major cause.

As one person explained:

Big fires come when that country is sick from nobody looking after with proper burning.

On a very pragmatic level, cleaning the country involves getting rid of long grass and grass seeds which impede travel. It means being able to see the animal tracks, and thus to hunt better. It means being able to see snakes and snake tracks so as to avoid them. Fire can be used to spread out the harvest of certain bush tucker over a long period of time.

April Bright explained:

Patterns of burning mean that certain areas are burnt at different times. This is important to the food chain. Smoke brings on flowering. For example, areas that are burnt early provide early hunting and foraging for both man and animal life. We follow the burns. For example, with the fruiting of, for instance, the apple trees and the plum trees. Those that have been burnt earlier, their fruiting comes on earlier, and as the fruit is on its way out in one place, the next patch of 'burn' will then produce plums and apples that can be picked.

Source: D.B. Rose, 1996, *Nourishing Terrains: Australian Aboriginal Views of Landscape and Wilderness*

a List the ways their use of fire was beneficial for the Aboriginal people.

____

____

____

b List the ways the Aboriginal use of fire was beneficial for the landscape.

____

____

____

c How might the introduction of fire-stick farming by Aboriginal people have altered the landscape?

____

____

2.4 Definitions: Landscapes and landforms

Knowledge and understanding

 **verbal–linguistic**

Write the correct term from the box for the definitions in the table below.

aesthetic value	constructed environment	Dreaming	place
landform	landscape	national identity	sacred sites
biophysical environment	multicultural		

Definition	Term
A key spiritual belief of Australian Aboriginal people; describes both the period of creation and the stories that come from this period	
The earth's human-altered landscapes; it includes all those features that are normally associated with settlements, industries and agriculture	
An area dominated by natural features such as landforms and vegetation; it includes the earth's soil, water, air, sunlight and all living things	
The value of a landscape based on its beauty or attractiveness	
Relating to several or more cultural or ethnic groups in a society	
A natural feature of the earth's surface; for example, a mountain, valley, lowland or volcano	
The overall appearance of an area resulting from the interaction of landforms, vegetation and soils together with human elements of the environment, such as transport networks, settlements, farms and factories	
The human and physical characteristics of a specific location on the earth's surface	
A person's sense of belonging to a nation	
Places of significant cultural and spiritual meaning to Aboriginal and Torres Strait Islander people	

2.5 The changing face of the earth

Geographical skills • Knowledge and understanding

mi visual–spatial • verbal–linguistic • logical–mathematical

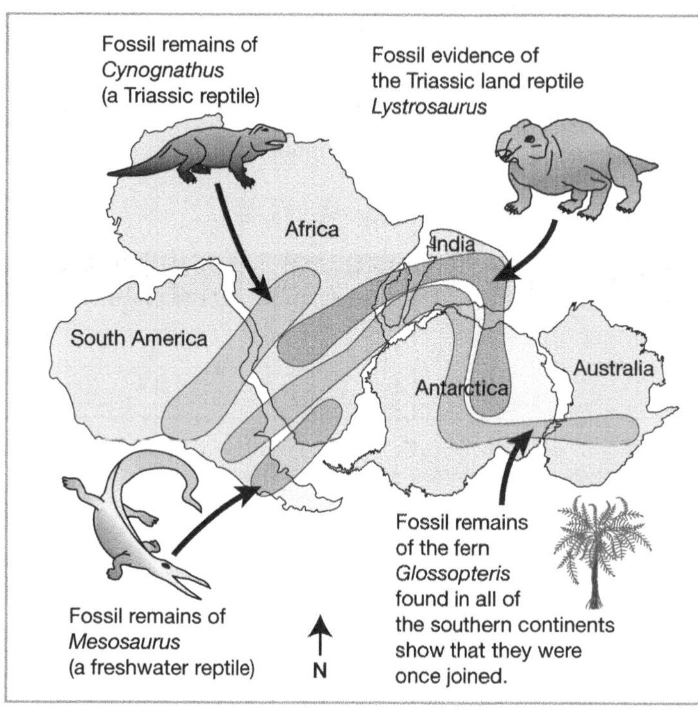

2.5.1

Fossil remains of reptiles as evidence for linking continents

1 Study the map in Figure 2.5.1. Name two continents which, according to fossil evidence, may have been connected.

____

2 Describe how the plates have been able to arrange themselves in their present locations.

____

____

____

____

3 Explain how fossils may be preserved within rock.

____

____

____

____

____

2.6 Weathering and erosion

Geographical skills • Knowledge and understanding

visual–spatial • verbal–linguistic

1 Use your knowledge of weathering to fill in the diagram in Figure 2.6.1.

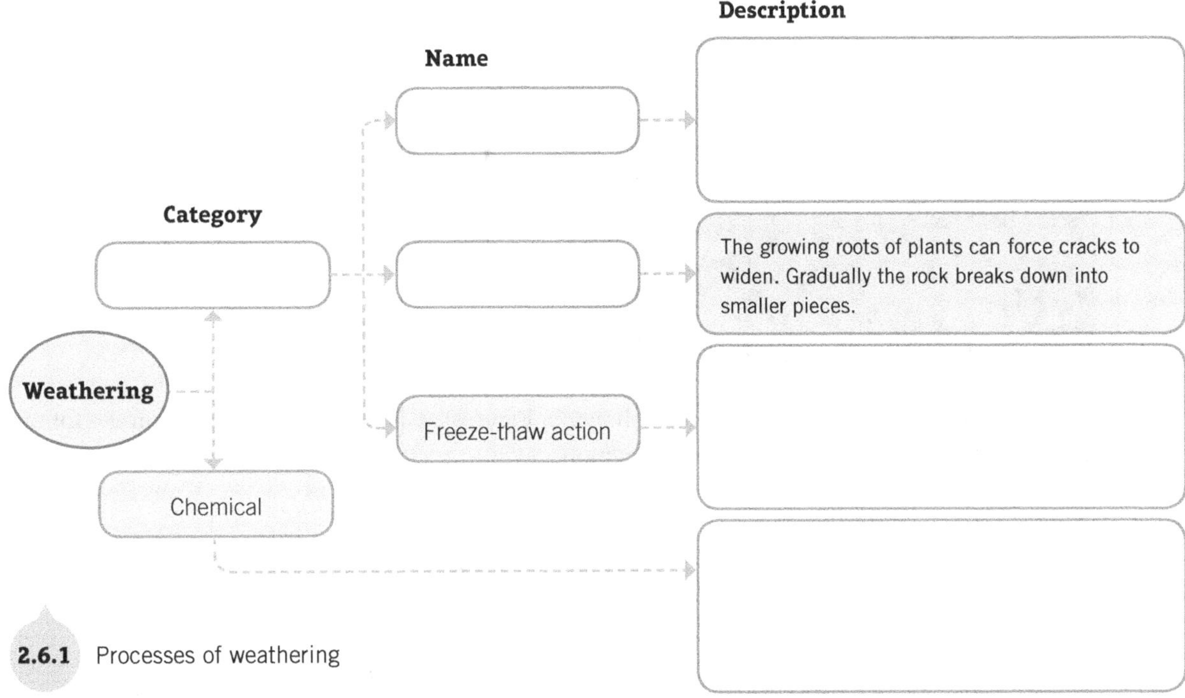

2.6.1 Processes of weathering

2 Identify and describe the processes of weathering in Figures 2.6.2, 2.6.3 and 2.6.4.

2.6.2 The type of weathering shown here is

____

Description: ______________________________

____

____

____

2.6.3 The type of weathering shown here is

____

Description: ______________________________

____

____

____

2.6.4 The type of weathering shown here is

____

Description: ______________________________

____

____

____

3 Identify and describe the type of erosion shown in Figures 2.6.5 and 2.6.6. Choose from the following: attrition, abrasion, corrosion or hydraulic action.

2.6.5 The type of erosion shown here is

____

Description: ______________________________

____

____

____

2.6.6 The type of erosion shown here is

____

Description: ______________________________

____

____

____

2.7 Coasts: Processes shaping landforms

Geographical skills • Knowledge and understanding

visual–spatial • verbal–linguistic

1 Explain the difference between an erosional coast and a depositional coast.

2 Use Figure 2.7.1 to explain why waves break.

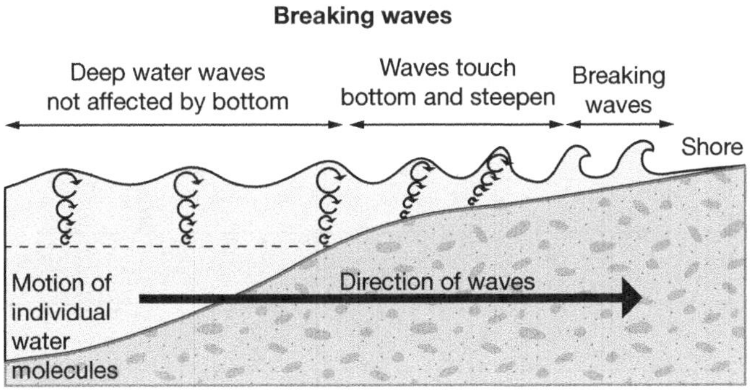

2.7.1 Waves breaking on a beach

3 Describe the process of swash and backwash, as illustrated in Figure 2.7.2.

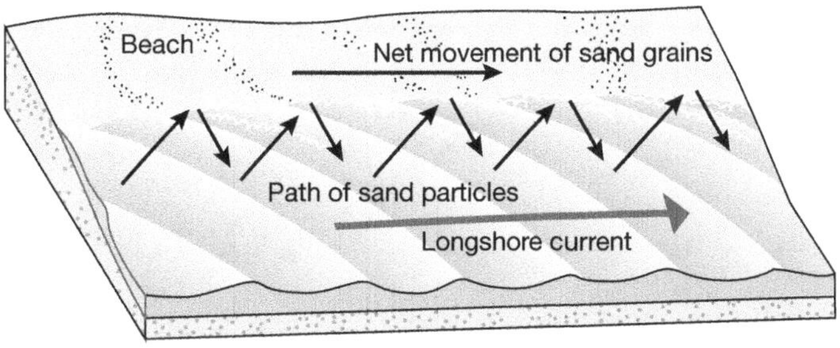

2.7.2 Swash and backwash along a beach

4 Study Figure 2.7.3.

2.7.3

Coastal cliff showing evidence of erosion

a What is erosion?

b Explain the appearance of the coastal cliff shown in Figure 2.7.3.

5 Draw a diagram of the moon and earth to explain why tides occur. Remember to include the phrase 'gravitational attraction'.

2.8 Features of coasts: Erosional and depositional landforms

Geographical skills • Knowledge and understanding

visual–spatial • verbal–linguistic

1 Study Figure 2.8.1. Identify three erosional landforms and label these on the illustration.

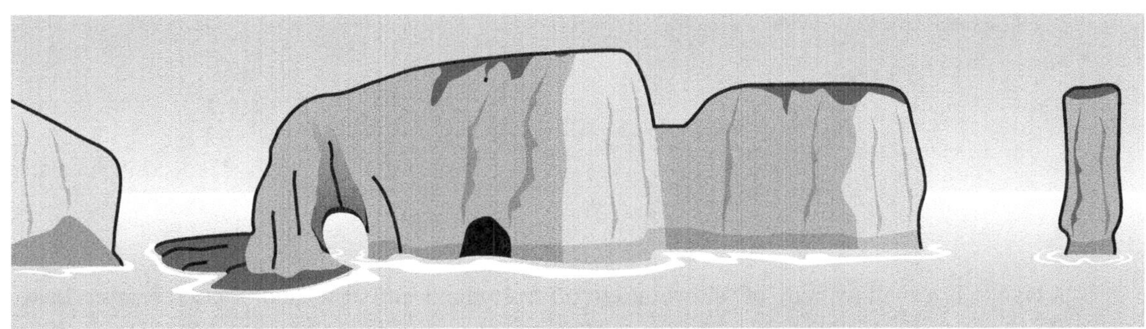

2.8.1 Coastal erosional landforms

2 Describe how each of the landforms in Figure 2.8.1 was formed.

a ______________________________

b ______________________________

c ______________________________

3 Study Figure 2.8.2.

Salt marsh created behind spit in low energy zone
Original coastline
Direction of longshore drift
Spit
Prevailing wind direction
Headland
Change in shape of land (initiation of deposition)
Spit stopped from growing any further due to river outlet removing material
Recurved laterals created by change in wind direction
Short-term change in wind direction

2.8.2 The formation of a spit

a Explain how the spit was formed.

____

____

____

b Explain how the salt marsh was able to form.

____

____

____

4 Study Figure 2.8.3.

a Label the diagram with the following: beach, tombolo, spit.

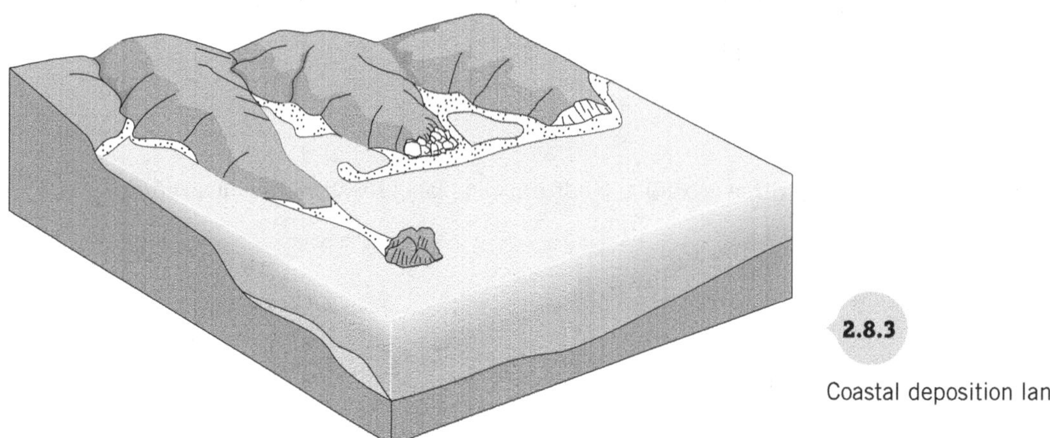

2.8.3

Coastal deposition landforms

b A lagoon is a shallow body of salt water cut off from the ocean by a sand or coral barrier. Based on this definition, label the lagoon in Figure 2.8.3.

5 Study Figure 2.8.4. Label the primary dune, the secondary dune and the back dune.

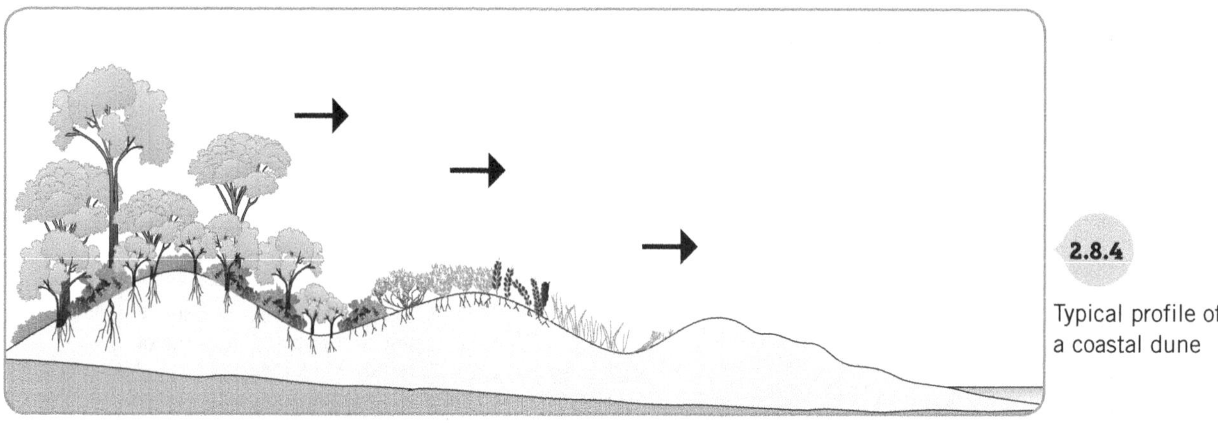

2.8.4

Typical profile of a coastal dune

6 Describe how a sand dune is stabilised.

____

____

____

7 Identify four ways a dune is affected by human activity.

a ____

b ____

c ____

d ____

2.9 Glacial landforms and processes

Knowledge and understanding • Geographical skills

mi visual–spatial • verbal–linguistic

Extension

1 Complete the following table by identifying the type of glacier and providing a brief description.

	Example of glacier	
a 2.9.1		**Type of glacier:** ________________ **Description:** ________________ ________________ ________________ ________________
b 2.9.2		**Type of glacier:** ________________ **Description:** ________________ ________________ ________________ ________________
c 2.9.3		**Type of glacier:** ________________ **Description:** ________________ ________________ ________________ ________________
d 2.9.4		**Type of glacier:** ________________ **Description:** ________________ ________________ ________________ ________________

2 Read the following information about icefields and answer the question that follows.

Icefields

Icefields are regions of ice found in cold climates and high altitudes. They are less than 50 000 km^2 and consist of interconnected valley glaciers from which nunataks, or higher peaks, rise.

The icefield is larger than alpine glaciers, smaller than ice sheets and similar in size to ice caps. The topography, or shape of the icefield, is determined by the surrounding landforms, unlike ice caps, which cover the landforms.

Explain how an icefield differs from the following:

a an ice cap: ____________________

b an ice sheet: ____________________

c a glacier: ____________________

3 Construct a meaningful paragraph about Juneau Icefield by numbering the following sentences in the correct sequence.

☐ Others, such as the Mendenhall Glacier, have retreated hundreds of metres.

☐ In all, there are 40 large and 100 smaller glaciers.

☐ Within this extensive area are many glaciers, such as the Taku Glacier and the Mendenhall Glacier.

☐ The ice field is the fifth largest in the Western Hemisphere and covers an area of 3900 square kilometres.

☐ Many tourists are attracted to the area.

☐ Located to the north of Juneau in Alaska is an ice field known as the Juneau Icefield.

☐ The Gilkey Glacier has retreated 3500 km from the furthest extent, known as the terminus of the glacier.

☐ Only one, the Taku Glacier, has been advancing.

☐ Since 1700, the ice field has been retreating.

☐ They are flown in by helicopter to walk on the ice and view the crevasses.

☐ Some of the ice is up to 1400 metres deep providing spectacular crevasses.

☐ Indeed, there are glaciers that have retreated thousands of metres.

4 Complete the table below with the correct type of glacier: niche glaciers, ice caps and ice sheets, cirque glaciers, valley glaciers, piedmont glaciers.

Type	Description
	Huge areas of ice covering more than 50000 sq km; while these once covered much of northern Europe and North America they are now found only in Antarctica and Greenland
	Form when valley glaciers extend onto lowland areas, spread out and merge
	Very small accumulations of glacial ice that are found in depressions and gullies
	Larger masses of ice that move down former river courses and are bounded by steep (almost vertical) valley sides
	Relatively small glaciers found in 'armchair-shaped' depressions in mountains

5 Three processes in a glacial landscape are erosion, transportation and deposition. Below are 10 statements about glacial processes. Label each statement as glacial erosion (E), glacial transportation (T), or glacial deposition (D).

- ☐ Ground moraine is the rock debris across a valley floor as a glacier retreats.
- ☐ Glacial abrasion is the scraping and grinding effect of angular material embedded in the glacier.
- ☐ After the glacial ice melts the valley formed by the smaller glacier is left 'hanging' high above the valley floor.
- ☐ Glacial hanging valleys form where a smaller tributary glacier joins a larger valley glacier and are often associated with waterfalls.
- ☐ Material is moved as debris carried within the body of the glacier.
- ☐ Material is moved on the surface of the glacier as lateral or medial moraine.
- ☐ Lateral moraine is debris formed as an embankment along the side of valleys as the glacier retreats.
- ☐ Glacial abrasion smooths and widens the U-shaped valley.

6a Draw arrows from the labels below to the appropriate part of the diagram to complete Figure 2.9.5.

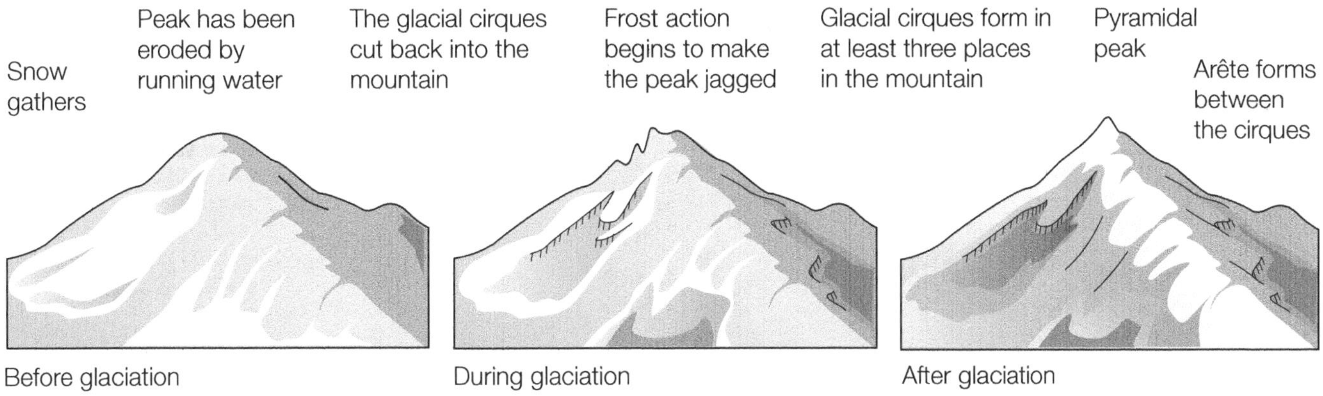

2.9.5 How a pyramidal peak is formed

b Use the completed diagram to explain how a pyramidal peak is formed.

____

____

____

7 In the formation of the pyramidal peak, identify a likely location for the process of glacial abrasion. What role does it play in the formation of a pyramidal peak?

____

____

____

8 Using Figure 2.9.6, explain how a hanging valley like the one in Figure 2.9.7 forms.

___

___

___

___

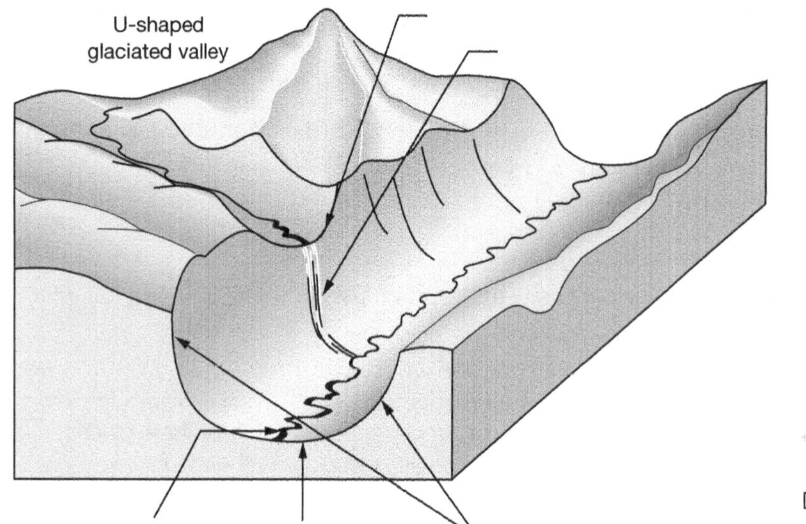

2.9.6

Diagram of U-shaped glaciated valley

2.9.7

A hanging valley

9 Complete the following process diagram to illustrate the formation of a tarn.

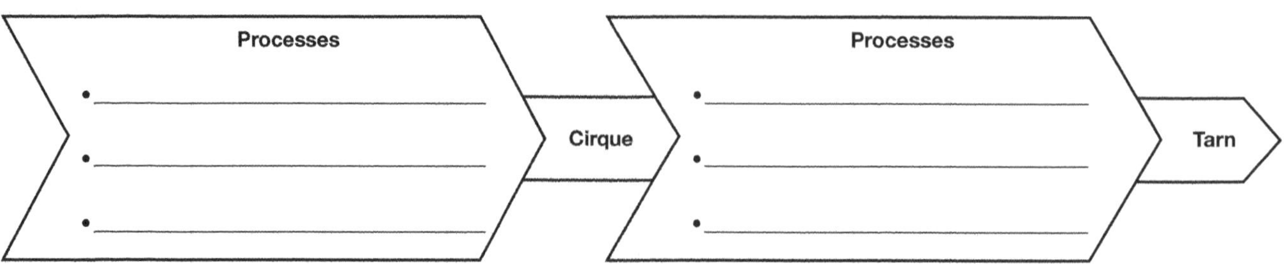

2.9.8 How a tarn is formed

2.10 Features of a glacier

Knowledge and understanding

verbal–linguistic • visual–spatial

1 Write the correct term for each of the following definitions:

Definition	Term
Unique to a defined geographic location	_ _ _ _ _ _ _ c
A long, narrow flooded valley with steep sides, carved by glaciation	_ j _ _ _
Slow-moving rivers of compacted snow and ice	_ _ _ c _ _ _
Rock material deposited by a glacier	m _ _ _ _ _ _
A lake at the base of a cirque	_ _ _ n
The sharp-sided ridge separating two cirques	_ _ e _ e
A small valley that is shaped like an amphitheatre	_ _ _ q _ _
Rock debris deposited across a valley floor as a glacier retreats	_ _ _ _ _ _ d m _ _ _ _ _ _ _ _
The rate at which the temperature of a body of air declines with elevation	l _ _ _ _ r _ _ _
The edge of the habitat in which trees are able to grow	_ _ _ e _ _ _ _
A valley with a flat floor and steep sides, formed by the abrasive power of material embedded in a glacier	_ -s _ _ _ _ _ v _ _ _ _ _

2 Help the following alpine terms 'ski' down to their correct chalet by drawing a line from each term to the correct definition.

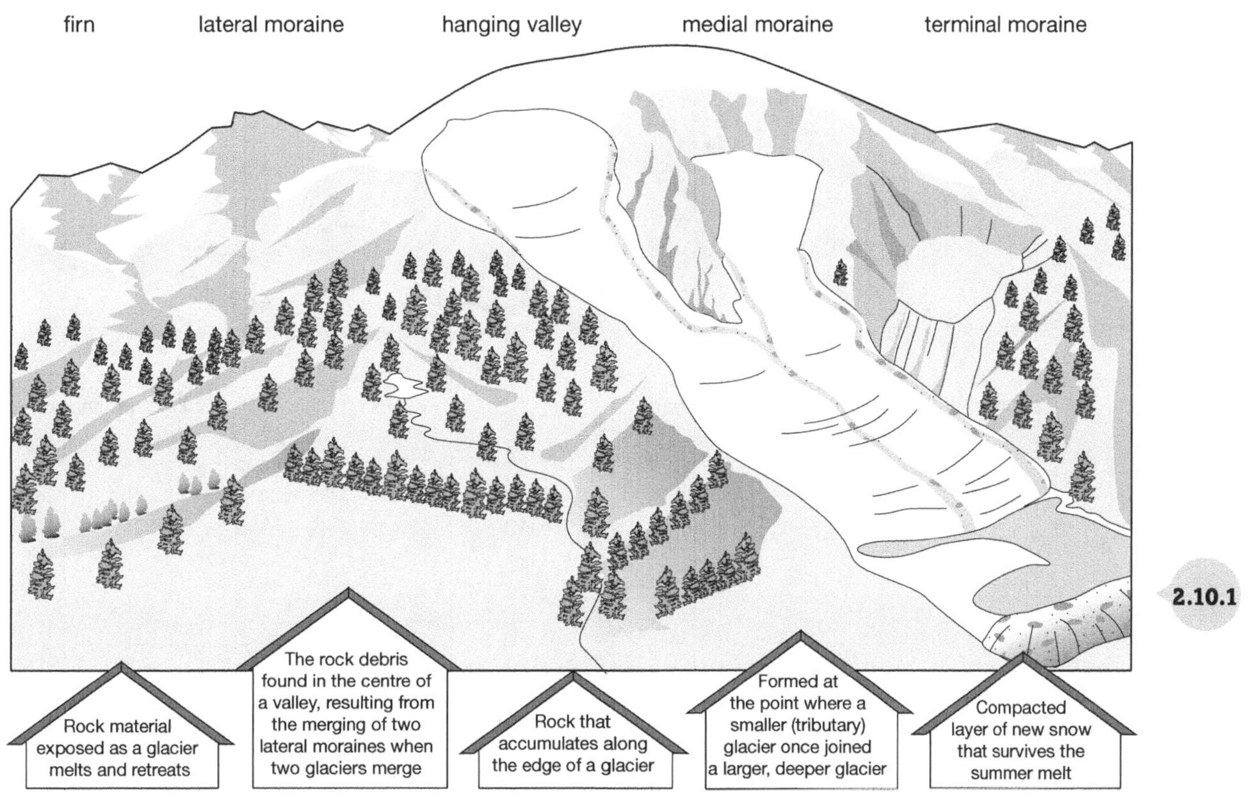

2.10.1

2.11 Features of rivers

Knowledge and understanding • Geographical skills

mi verbal–linguistic • visual–spatial • logical–mathematical

1 Label each of the images below with the terms above them.

Interlocking spurs | Watershed | River channel

2.11.1 V-shaped valley

Escarpment | Plunge pool | River

2.11.2 Waterfall

Braided stream river | Floodplain | Temporary islands

2.11.3 Floodplain with braided stream

River channel | Meander | Floodplain

2.11.4 Meanders

2 Niagara Falls (USA and Canada) is one of the world's most recognisable waterfalls. This iconic waterfall attracts tourists from around the world.

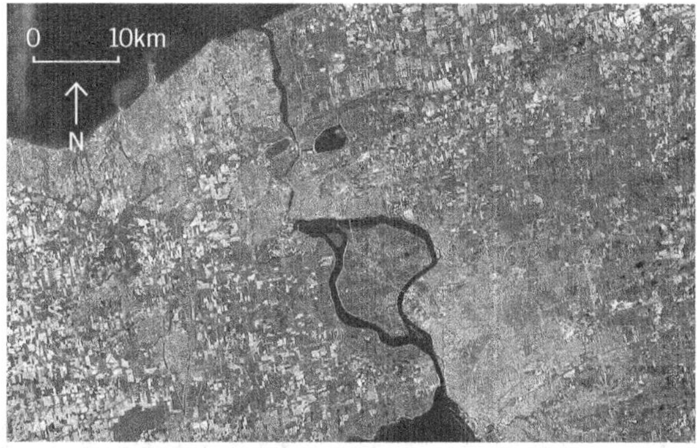

2.11.5 Satellite image of Niagara Falls region

2.11.6 Oblique photo of Niagara Falls

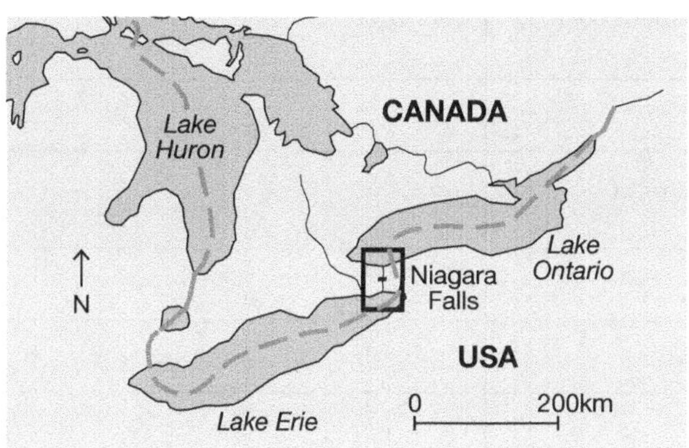

2.11.7

Map of Niagara Falls region

a What is the length of the Niagara River? ______________________

b Name the lakes that the Niagara River flows between. ______________________

____

c In which direction does the river flow? ______________________

____

d Describe the pathway of the Niagara River.

____

____

____

e In which direction was the photographer facing when they took the photo in Figure 2.11.6?

____

2.12 River features

Knowledge and understanding • Geographical skills

verbal–linguistic

1 Write the correct term from the box for the definitions in the table below.

abrasion	bed load	braided channel	delta	deposition
infiltration	suspended load	tributary	turbidity	watershed

Definition	Term
The movement of water from the land surface into the soil	
Extensive deposits of alluvial material (sediment) at the mouth of a river	
The boundary between catchments	
Muddiness; high turbidity occurs when there are high levels of suspended sediment in water	
A river system flowing into a larger river	
The wearing down or wearing away of rock by friction	
Material transported along the bed of a river	
Fine particles of silt and clay carried in river water	
Accumulation of sediment by the action of erosional agents, such as water and wind	
A river channel featuring a network of small channels separated by small and often temporary islands of sediment	

2 Highlight the following words, in green for 'deposition' and in red for 'erosion'.

waterfall	floodplain	levee bank
point bar	meander	delta
ox-bow lake	rapids	sediment islands

2.13 Desert landforms

Knowledge and understanding • Geographical skills

 verbal–linguistic • visual–spatial

1 a Label the photos below with one of the four main types of dunes.

b Suggest how each may have formed based on the patterns shown.

2.13.1

Type of dune: ______

______

______

2.13.2

Type of dune: ______

______

______

2.13.3

Type of dune: ______

______

______

2.13.4

Type of dune: ______

______

______

2 Which do you consider the most powerful force of erosion in desert regions, wind or water? Explain your answer.

______

______

______

Extension The formation of Uluṟu

Uluṟu is an internationally famous landform found in the desert of Central Australia. Travelling towards Uluṟu, it looks smooth and featureless but up close it is pitted with holes, ribs and caves.

3 Read the explanation of the formation of Uluṟu. Write the number for the correct caption beside the diagrams.

A 550 million years ago, the Peterman Ranges were as tall as the Himalayas are now. Rainwater flowing down the mountains eroded sand and dropped it on the surrounding plain and it became kilometres thick.

B 500 million years ago, the whole area became covered by an inland sea. Sand and mud fell to the bottom of the sea and covered the seabed. The weight of the new seabed turned the material beneath into rock. The sand was compressed to form Arkose sandstone.

C 400 million years ago, the sea disappeared and massive forces folded and tilted the sandstone to 90 degrees so the layers of sandstone became vertical.

D 30 000 years ago windblown sand formed dunes and plains. Uluṟu is the tip of a huge slab of rock that lies below the ground for possibly 5 to 6 kilometres.

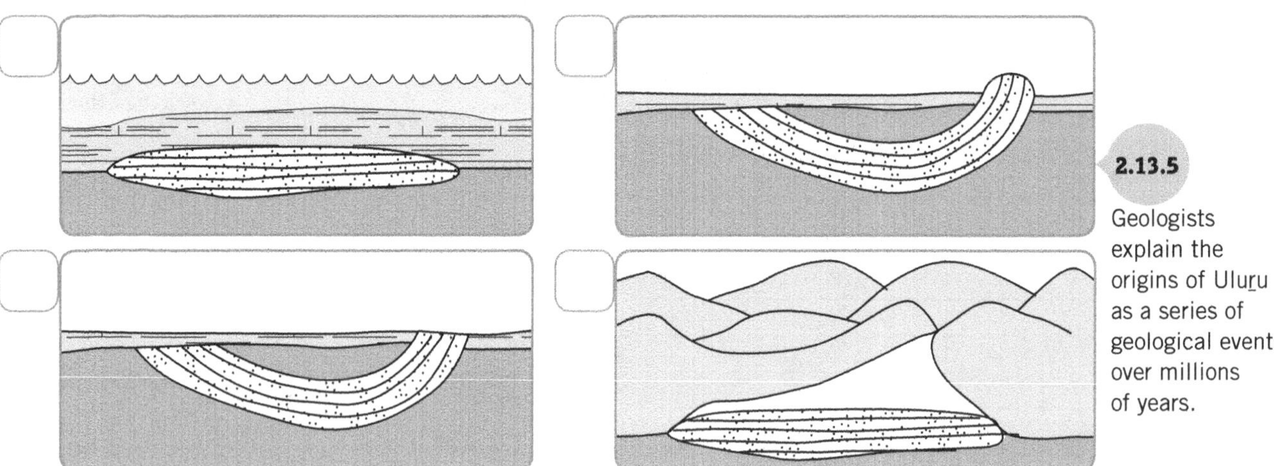

2.13.5 Geologists explain the origins of Uluṟu as a series of geological events over millions of years.

Traditional owners

The geological name for this landform is an 'inselberg'. But to the A<u>n</u>angu People, the traditional owners, Uluṟu is part of their 'Tjukurpa', the stories that explain the existence of life and guide daily activities.

4 Research online and find a version of A<u>n</u>angu 'Tjukurpa' that explains how Uluṟu came to be. Write it out in point form in the space below.

______

______

______

______

______

______

5 Which version of the formation of Uluṟu do you prefer? Why?

______

______

2.14 Features of a desert

Knowledge and understanding • Geographical skills

 verbal–linguistic • visual–spatial

1 Look at the photos below. Use the hints to help you name and describe these desert features.

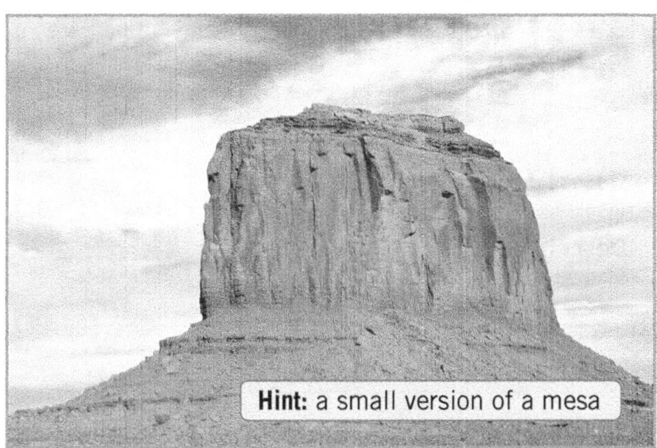

Hint: a small version of a mesa

2.14.1

Desert feature: ____________________

Description: ____________________

____

____

____

Hint: can live in very dry areas

2.14.2

Desert feature: ____________________

Description: ____________________

____

____

____

Hint: a place to get refreshed in the desert

2.14.3

Desert feature: ____________________

Description: ____________________

____

____

____

Hint: nothing but sand

2.14.4

Desert feature: ____________________

Description: ____________________

____

____

____

2.14.5

Desert feature: ______________________

Description: ______________________

____

____

2.14.6

Desert feature: ______________________

Description: ______________________

____

____

____

Hint: I have only a very short life when it rains

2.14.7

Desert feature: ______________________

Description: ______________________

____

____

____

2 In the table below, write the correct term with each definition: aquifer, bolson, desertification, erg, inselberg, reg, wadi.

Definition	Term
A large erosion-resistant feature surrounded by an eroded plain	
An inland desert basin surrounded by mountains	
The process by which productive land in arid areas is changed into desert	
A desert surface covered with sand dunes	
A layer of permeable rock that is capable of storing significant quantities of water	
A stone-covered desert	
A watercourse or gully that is dry except during periods of rainfall	

3.1 Land clearance in Australia

Knowledge and understanding • Geographical skills

visual–spatial • logical–mathematical

Study the statistics on land clearance and complete the tasks below.

Rates of land clearing in Australian states and territories

State/Territory	Percentage %	Degrees	Key (colour)
Queensland	75		
New South Wales	16		
Tasmania	2		
Northern Territory	2		
Western Australia	2		
Victoria	2		
South Australia	1		

Source: ABS

1 Complete the pie graph using the following steps.

a Calculate the degrees representing the percentages of land clearance using the formula 1% = 3.6°.

b Add your results to the table.

c Using a protractor, complete the divisions for the pie graph.

d Select a range of colours for the key and use these colours to shade the divisions on your graph.

e Use the same colours to shade the map of Australia (Figure 3.1.2) to show land clearance.

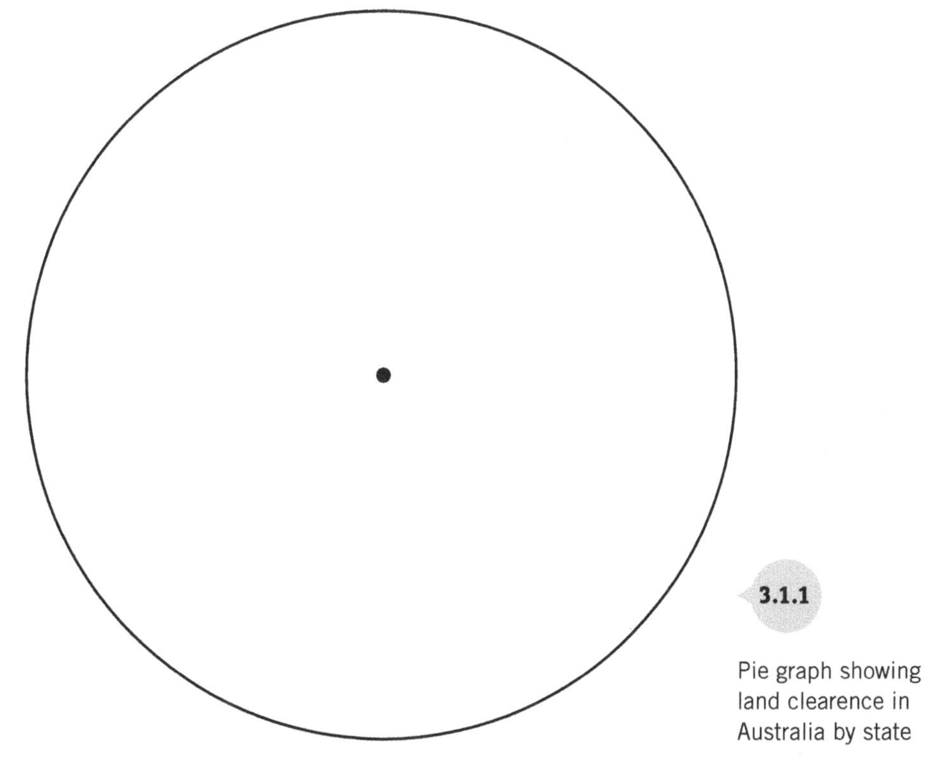

3.1.1

Pie graph showing land clearence in Australia by state

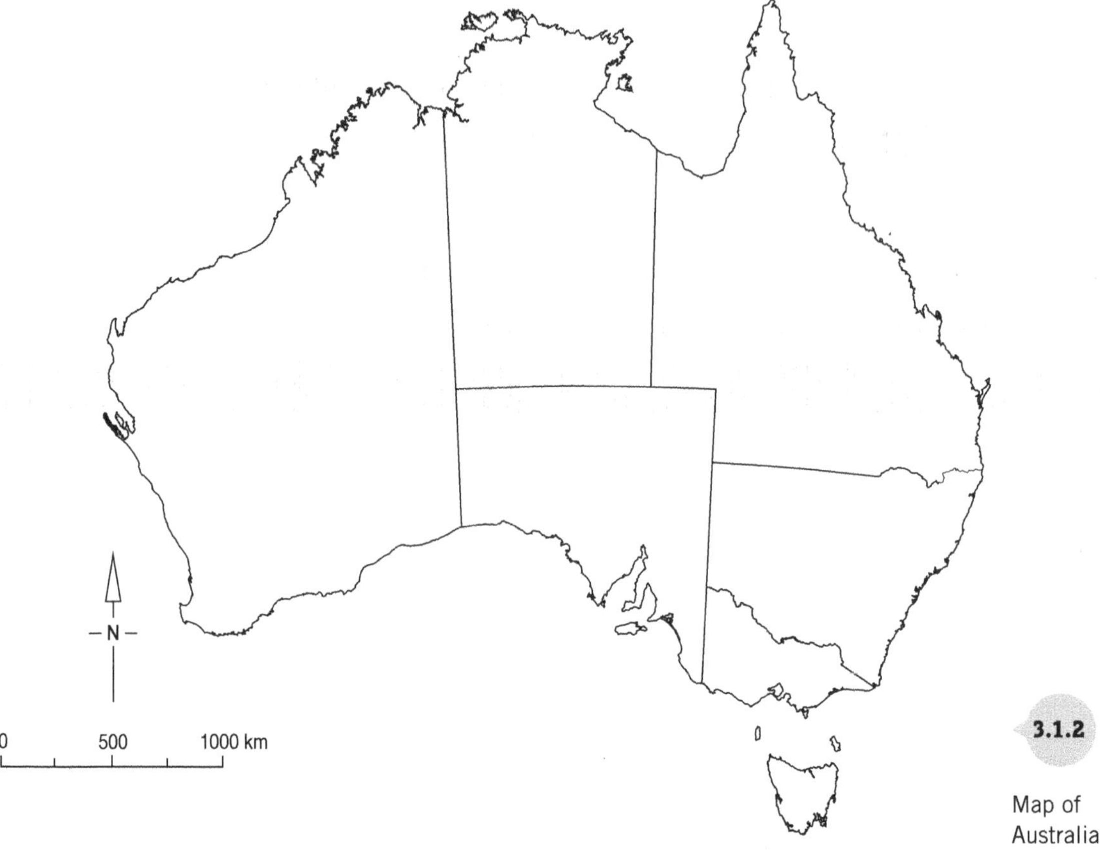

3.1.2

Map of Australia

2 Use your pie graph and map to answer the following questions.

a Which region of Australia had the highest percentage of land clearance?

____

b Which region of Australia had the lowest percentage of land clearance?

____

c Which state has the highest percentage of land clearance?

____

d Which state has the lowest percentage of land clearance?

____

e From the list below, choose the factor that you think is the most important in relation to land clearance. Explain your choice.

- historical development
- existing vegetation cover
- population
- agriculture/cropping
- topography (shape of the land)
- forestry/wood products
- urban growth

____

____

____

3.2 Soil erosion

Knowledge and understanding

 visual–spatial • interpersonal

Types of erosion

1 Label each photo with the type of erosion shown.

3.2.1 ____________________

3.2.2 ____________________

3.2.3 ____________________

3.2.4 ____________________

2 Label the following diagram to show the four main types of soil erosion.

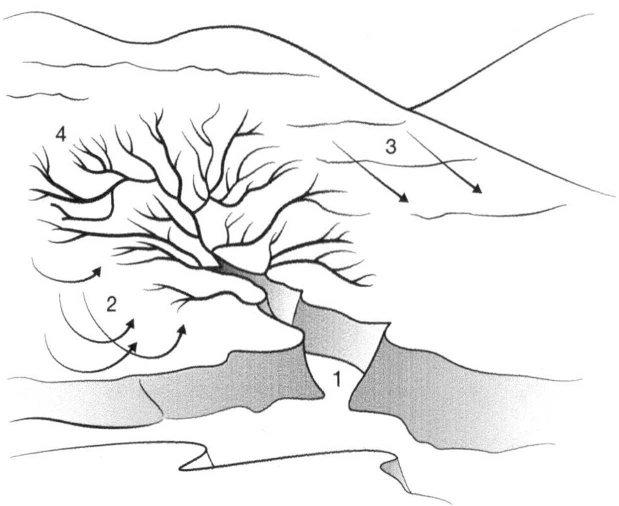

3.2.5

Four main types of soil erosion

3 Imagine you are TV reporter. Select one of the types of erosion shown in Figures 3.2.1 to 3.2.4. Prepare a script about the event that causes the erosion. Include the following:

- a description of the event at the scene
- a report by a meteorologist on the climate and weather factors contributing to the event
- an interview with an expert soil scientist on the physical environmental factors causing this event.

3.3 Australia's national parks

Knowledge and understanding • Geographical skills

 **verbal–linguistic • visual–spatial • logical–mathematical**

Please don't climb Uluru

We, the Anangu traditional owners of Uluru-Kata Tjuta National Park, have a responsibility to teach and safeguard visitors to our land. We feel great sadness when a person dies or is hurt on our land. We would like to educate people on the reasons we ask you not to climb and if you choose to climb, we ask that you do so safely.

Cultural reasons

We ask visitors not to climb Uluru because of its spiritual significance as the traditional route of the ancestral Mala men on their arrival at Uluru. We prefer that visitors explore Uluru through the wide range of guided walks and interpretive attractions on offer in the Park. At the Cultural Centre you will learn more about these and the significance of Uluru in Anangu culture.

Safety reasons

The climb is physically demanding and can be dangerous. More than 35 people have died while attempting to climb Uluru and many others have been injured. At 346 metres, Uluru is higher than the Eiffel Tower or as high as a 95-storey building. The climb is very steep and can be very slippery. It can be very hot at any time of the year and wind gusts can hit the summit or slopes at any time. Every year people are rescued by park rangers, many suffering serious injuries such as broken bones, heat exhaustion and extreme dehydration.

Environmental reasons

There are also significant environmental impacts of climbing Uluru. If you have a close look you can see the path is smooth from thousands of footsteps since the 1950s. This erosion is changing the face of Uluru.

Also, there are no toilet facilities on top of Uluru, and no soil to dig a hole. When it rains, everything gets washed off the rock and into the waterholes where precious reptiles, birds, animals and frogs live. A water quality study at Uluru has found significantly higher bacterial levels in the waterholes fed by runoff from the climb site, compared with those further away.

Source: Australian Government Director of National Parks website

1 Why is Uluru important to the Anangu?

____

____

2 Why would the Anangu prefer visitors not climb Uluru?

____

____

3 What other reasons are given for not climbing Uluru? Can you think of any others?

____

____

____

4 Why would some visitors still want to climb Uluru? ______________________________

____

____

____

3.4 Managing coasts

Geographical skills • Knowledge and understanding

 visual–spatial • verbal–linguistic

1 Study Figure 3.4.1 then answer the questions that follow.

a Identify the type of photograph in Figure 3.4.1: ground level, aerial or oblique?

____

b Identify the coastal deposition landform in Figure 3.4.1.

____

c Draw an arrow to show the direction of longshore drift. Explain how you were able to determine the direction.

____

____

____

____

3.4.1 Beach management of a coastal depositional landform

d Identify two beach management strategies and explain how they preserve the beach.

____

____

____

____

2 Study Figure 3.4.2 then answer the questions that follow.

a What does Figure 3.4.2 show?

____

b Shade in the area affected on the photograph in Figure 3.4.2.

3.4.2

c Explain how the blowout may have occurred. ______

______

______

d Suggest two strategies that could be implemented to manage this dune blowout.

______

______

______

______

3 Read the article below about Kingscliff Beach then answer the questions that follow.

3.4.3

Aerial photographs of Kingscliff in New South Wales on Australia's east coast, seen from the air in 2006 (A), and in 2012 (B) without its expanse of sand

Australia's surfers mourn disappearing east coast beaches as currents sweep sand out to sea

Hundreds of kilometres of Australia's most popular beaches are shrinking as ocean currents sweep their sands out to sea, to the dismay of millions of surfers.

At Kingscliff Beach on Australia's east coast, home to one of the country's oldest surf clubs, tourists and locals have long spent summers on the lush stretch of golden sand. But now the beach is no longer there.

Instead, the 65-yard-wide stretch of sand – hundreds of thousands of cubic metres of it – has disappeared, swept into waters to the north where it has been carried by the tide to form sandbanks.

And with it have gone some of the classic images of Australia's coast – a surfing mecca that lures more than two million people a year to ride the waves, and many more to swim off the spectacular beaches.

The rapid disappearance of the sand has puzzled scientists, who are still trying to gauge its varying causes and are examining whether – and to what extent – the beaches will ever be naturally restored by currents bringing new deposits of sand from elsewhere.

The most likely culprit, they believe, is a change in weather conditions, with increasingly frequent storms caused by La Niña, a cyclical cooling in ocean temperatures that seems to be recurring more frequently.

Along the coast, powerful storms and strong tides have swept away the sand, while changes in wave direction have dragged it offshore. As a result there is a new threat to coastal properties, with erosion of cliffs accelerated not because of rising tides—the ocean at Kingscliff only has a tidal variation of about five feet—but because the beach that dissipated the power of the waves has been diminished as the sand has gone.

Scientists say it could take a decade or more for these beaches to be naturally restored—if, indeed, they ever are.

'This is a new situation for us–the sand is all gone,' said Richard Adams, who runs a holiday park overlooking what used to be the beach at Kingscliff, 50 miles [80 km] south of Brisbane. 'This was all thick fluffy white sand, not this white water you can see now. It used to be one of the best beaches in Australia.'

'We had pretty calm weather throughout the last three decades—now we are moving back into an era of stormy conditions, especially with La Niña,' Prof Tomlinson said. 'The erosion seems to be triggered by reasonably sized storms and very subtle shifts in wave direction.'

The disappearance of the sand has worried local lifesavers, who stand trained and ready with rescue equipment but have little or no beach to patrol.

'We're a bunch of lifesavers who essentially can't get on to our beach,' said a disgruntled team member, Andrew Jones. His beach, Old Bar in northern New South Wales, has lost 75 yards [69 metres] of frontage in the past 18 months. A report commissioned by Surf Life Saving Australia, the organisation responsible for water safety and rescue, found 63 per cent of the country's surf lifesaving clubs were themselves erected in 'zones of potential instability'.

At Kingscliff, where a hastily built wall has helped save the headquarters from collapse, locals joke that they may have to turn their 90-year-old surf club into a yacht club.

...

According to Prof Tomlinson, 'Australians ... have not really adjusted. We still have that special relationship to the beach; it's our holiday destination, our economy, our relaxation. But in the absence of proper management, we are going to see a long term reduction in our beaches.'

Source: Jonathan Pearlman 2012, *The Telegraph*, Kingscliff

a Draw a line on the 2006 photograph of Figure 3.4.3 showing where the shoreline is in 2012.

b Compare the photographs of Kingscliff Beach and list differences between the two dates.

c Explain how coastal processes affect Kingscliff Beach.

d From the photograph, identify and describe beach management strategies in 2006.

e Explain why you think it is possible or not possible to restore the beach to the way it was in 2006.

4.1 Definitions: Plate tectonics

Knowledge and understanding

 verbal–linguistic

Write the correct term for the definitions in the table below. Select from the terms in the box. Not all terms need to be used.

block mountain	crust	deposition	differential erosion	earthquake
fold mountain	erosion	exfoliation	faulting	folding
mid-ocean ridge	mantle	ocean trench	plate	tectonic plate
margins	rift valley	seafloor spreading	vertical exaggeration	volcano
weathering				

Definition	Correct term
The separation of oceanic plates	
A large elongated depression with steep walls, formed by the downward movement of a block of the earth's surface between nearly parallel faults	
The thin outer layer of the earth (the lithosphere)	
The physical or chemical breakdown of rocks into smaller pieces	
A large segment of the earth's crust that is slowly moving due to convection currents in the mantle	
A sudden movement of the earth's crust caused by the release of pressure	
An underwater depression created when oceanic plates are drawn down into the earth's mantle	
The buckling of rock due to pressure	
An underwater ridge formed when continental plates move apart, allowing molten material to fill the gap created	
The wearing away of softer rock at a different (faster) rate than harder, more resistant rock	
The fracturing of rock along lines of physical weakness	
The wearing down, transportation and deposition of material by water, wind and ice	
The process of rock breakdown that occurs when rock expands and contracts due to temperature changes, then cracks, peeling off in layers like an onion	
The layer between the earth's core and its crust	

4.2 Earthquakes

Geographical skills • Knowledge and understanding

 **visual–spatial • verbal–linguistic**

1 Explain why shockwaves occur during an earthquake.

____

____

2 Determine if the following effects of an earthquake are primary or secondary.

Effect of an earthquake	Primary or secondary effect
Humans crushed to death	
Fear among people	
Interrupted phone connections	
Fires caused by broken gas pipes	
Collapsed buildings	
Loss of electricity supply	
Broken sewerage pipes	
Roads destroyed	
People injured	

Earthquake in Sichuan, China

At 2:28 pm, on Monday 12 May 2008 a magnitude 7.8 earthquake occurred in the Sichuan province of China. The epicentre was in Wenchuan County, 80 km west-north-west of Chengdu. The earthquake was felt 1500 km away in Beijing. This was partly due to the firm land surrounding the epicentre. As a result, the earthquake was able to impact a very large area. At the time, population density was high – there were 15 million people living in the affected area. In total, 69 197 people were killed, 18 222 listed as missing and 374 176 injured. Many of those killed were schoolchildren, trapped in poorly built school buildings.

Approximately 4.8 million people were made homeless, although some reports put the number of homeless at 11 million. Near the epicentre, almost 80% of the buildings were destroyed. The damage to buildings was mostly due to the fact that the buildings tended to be older and built prior to 1976. After a massive earthquake in 1976 China brought in special building regulations to ensure buildings could withstand the impact of earthquakes. But even new buildings had not been constructed to approved designs since they were in poorer rural communities.

Heavy rain caused landslides, blocking rescue efforts as emergency vehicles made their way to assist.

Source: Adapted from Wikipedia (accessed 29 July 2013)

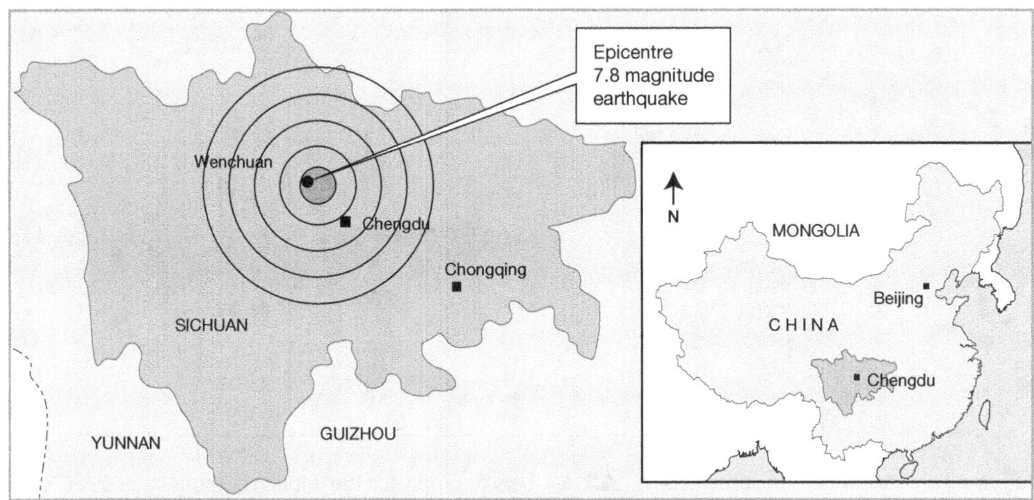

4.2.1

Map showing the location of the 2008 earthquake in Sichuan, China

4.2.2

Damage caused by the 2008 earthquake in Sichuan, China

3 Explain how the following factors may have influenced the impact of the 2008 earthquake in Sichuan (Figures 4.2.1 and 4.2.2).

Factor	Influences on the extent of damage and loss of life following an earthquake
Distance from the epicentre	
Level of preparation	
Population density	
Time of day	
Season – including weather and climate	
Type of land on which settlements are built	

4.2.3 The tsunami following an earthquake in Japan 2011

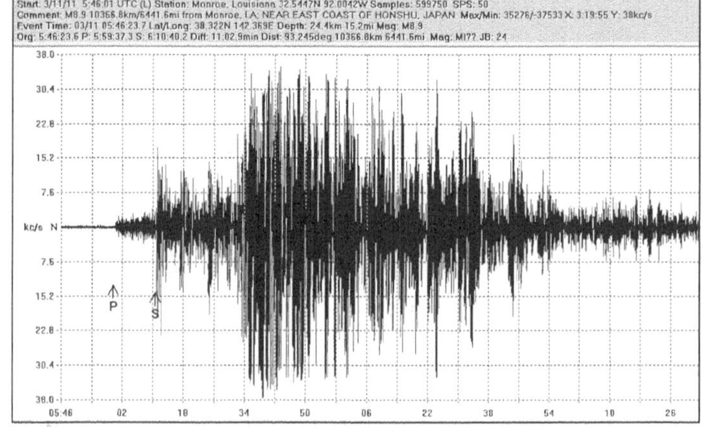

4.2.4 Seismograph for the Japan earthquake at 2:46 pm local time on 11 March 2011

4 What is a tsunami?

____

5 Describe how the location in Figure 4.2.3 is changing as a result of the tsunami.

____

____

6 Explain how an earthquake can result in a tsunami.

____

____

____

Refer to Figure 4.2.4 to answer questions 7–9.

7 Describe how a seismograph works.

____

____

8 What three main things does a seismograph measure?

a ____

b ____

c ____

9 a At precisely what time did the earthquake begin?

____

b At what time were the largest shockwaves felt? ______________________

____

c How long did the main earthquake last? ______________________

____

4.3 Volcanoes

Geographical skills • Knowledge and understanding

 **visual–spatial • verbal–linguistic**

1 What determines the type of volcanic eruption (explosive or not)?

____

____

2 Name and describe the following types of volcanoes.

4.3.1 Mt Fuji, Japan

Type of volcano: ______________________

Description: ______________________

____

____

____

4.3.2 Mauna Loa, Hawaii, USA

Type of volcano: ______________________

Description: ______________________

____

____

____

4.3.3 Kaguyak volcano, Alaska, USA

Type of volcano: ______________________

Description: ______________________

____

____

____

Read the article below then answer the questions that follow.

Australian volcano starts to blow

Don't worry: only penguins are in danger as it's in the middle of nowhere.

One of Australia's two active volcanoes seems to be erupting.

We say *seems* because the volcano in question, on Heard Island, is located in the southern reaches of the Indian Ocean, 2000 km north of Antarctica and closer to Africa than to Australia. That's about as close to the middle of nowhere as it is possible to be.

Heard Island and its neighbour MacDonald Island are Australian territories and are uninhabited, but each possesses an active volcano. Scientific expeditions venture there infrequently, due to conservation issues and the fact the islands have a wretched climate, are thousands of miles from anywhere nice and can only be accessed or supplied by ship.

That means little attention is paid to the islands, with the satellite images seldom acquired.

But NASA's Earth Observatory says and other analysis have detected heat signatures on Heard Island's 2475 m Mawson Peak that suggest recent volcanic activity.

'Although not definitive, this natural-color satellite image also suggests an ongoing eruption,' NASA writes. 'The dark summit crater (much darker than Mawson's shaded south-western face) is at least partially snow-free, and there's a faint hint of an even darker area – perhaps a lava flow – within. Shortwave infrared data (collected along with the visible imagery) shows hot surfaces within the crater, indicating the presence of lava in, or just beneath, the crater.'

Source: Simon Sharwood, *The Register*, 24 October 2012

3 Describe the volcanic activity on Heard Island.

4 Account for the volcanic activity on Heard Island.

5 How does the volcanic activity on Heard Island differ from activity on the Australian mainland?

6 Why do you think this is so?

Read the article below then answer the question that follows.

Scores killed by eruption from Indonesian volcano

A further 54 bodies were brought in after the latest eruption of Mt Merapi and more than 66 others were injured, many of them critically with burns.

The latest deaths bring the toll to more than 90 since the country's most active volcano started erupting on October 26.

Many of the dead were children from Argomulyo village, 18 kilometres from the crater of the volcano, according to witnesses and emergency response officials.

"Argomulyo village has been burned to the ground by the heat clouds. Many children have died there. When I was in the village the ground was still hot," a Yogyakarta police force medic said.

A river running through the village overflowed with a thick mixture of mud and ash, and several bodies lay unclaimed in the debris, witnesses said.

Ash, deadly heat clouds and molten debris gushed from the mouth of the mountain which stands about 3000 metres, and shot high into the sky for most of the night and into the morning.

There was panic and chaos on the roads as people tried to flee in the darkness, rescue workers said.

The numbers of evacuees swelled past 100 000 people, with 30 000 moved into a sports stadium away from the peak.

"The emergency shelters are now overcrowded," an emergency response field coordinator said.

The international airport at Yogyakarta was closed as ash clouds billowed to the altitude of cruising aeroplanes and the runway was covered in grey soot.

Surono, a government volcanologist, said the blasts were the largest yet.

"This is the biggest eruption so far. The heat clouds went down the slopes as far as 13 kilometres and the explosion was heard as far as 20 kilometres away," he said.

The exclusion zone around the mountain was widened and people living in the area were evacuated.

Indonesia's transport ministry has told pilots to stay at least 14 kilometres away from the volcano, and several flights between central Java and Singapore and Malaysia have been cancelled this week.

President Susilo Bambang Yudhoyono visited people displaced by the volcano as the disaster-prone country struggled to cope with a second natural disaster following the tsunami off Sumatra on October 25.

Source: Adapted from *The Telegraph, 5 November 2010*

7 Compare the eruption on Mt Merapi, Indonesia, with the eruption on Heard Island, Australia.

4.4 Features of a volcano

Knowledge and understanding

 **verbal–linguistic**

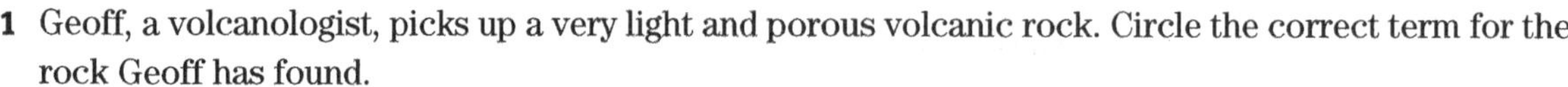

1 Geoff, a volcanologist, picks up a very light and porous volcanic rock. Circle the correct term for the rock Geoff has found.

caldera lahars maar pumice scoria

2 Draw a line to connect the following terms with their definition.

Term	Definition
focus	Location where hot molten magma from deep within the earth rises up through the crust to reach the surface
volcanism	A volcanic landform feature formed when a violent eruption blasts away the top of an existing volcanic cone or shield
lahar	Processes associated with volcanoes and volcanic activity
epicentre	A volcano that is said to be 'sleeping', as it has not erupted for a period of time but may still erupt
dormant volcano	A mudflow formed when volcanic material mixes with water
caldera	A volcano that has not erupted for a long period and will not erupt again
extinct volcano	A point under the earth's surface at which a sudden movement in the earth's crust occurs; the origin of an earthquake
hot spot	A point on the earth's surface that is directly above the centre (focus) of an earthquake

3 Draw a volcano in the space below, including and labelling as many of the terms listed above as you can. You may like to add other terms too.

5.1 Liveability of places

Knowledge and understanding • Geographical skills

 **verbal–linguistic • visual–spatial**

A tag cloud (or word cloud) is a visual representation of the number of times a word is used in a text. In the example in Figure 5.1.1, residents of San Francisco in the United States were asked to describe their idea of liveability. The words they used were recorded and turned into a tag cloud. The size of a word in the tag cloud reflects the frequency with which it was mentioned. The larger a word appears in the tag cloud, the more frequently it was mentioned and the more important it is in defining liveability for the people interviewed.

5.1.1

A tag cloud of liveability in San Francisco, United States

1 Use the terms in Figure 5.1.1 to write your own definition of liveability.

____

____

____

____

2 Consider your suburb or town. List three features specific to where you live and explain how each contributes to or improves liveability.

____

____

____

____

____

____

____

____

5.1

The liveability of urban areas can be measured against a number of criteria. A city can be measured using these criteria and given a score. The criteria can also be used to see how a city could be improved. Some liveability criteria are shown in Table 5.1.2.

Liveability criteria	Your nearest city
Amount of green space	
Amount of urban sprawl	
Presence of natural features	
Presence of history and culture	
Quality of public transport	
Levels of pollution	
Affordable housing	
Quality of public healthcare	
Quality of education	
Quality of roads	

5.1.2 Liveability criteria for your nearest city

3 Think about the city that is nearest to where you live. Complete Table 5.1.2 by giving that city a score for each criteria ranging from 1 (lowest) to 10 (highest).

4 Pick one criteria and explain how the city could be improved in this area.

____

____

____

5 Think of other criteria you could use to assess how liveable a city is. List six criteria you would consider if you were making your own liveability index.

- ______________________________________
- ______________________________________
- ______________________________________
- ______________________________________
- ______________________________________
- ______________________________________

5.2 Measuring liveability

Knowledge and understanding

verbal–linguistic • visual–spatial

Study the map in Figure 5.2.1 and then answer the questions that follow.

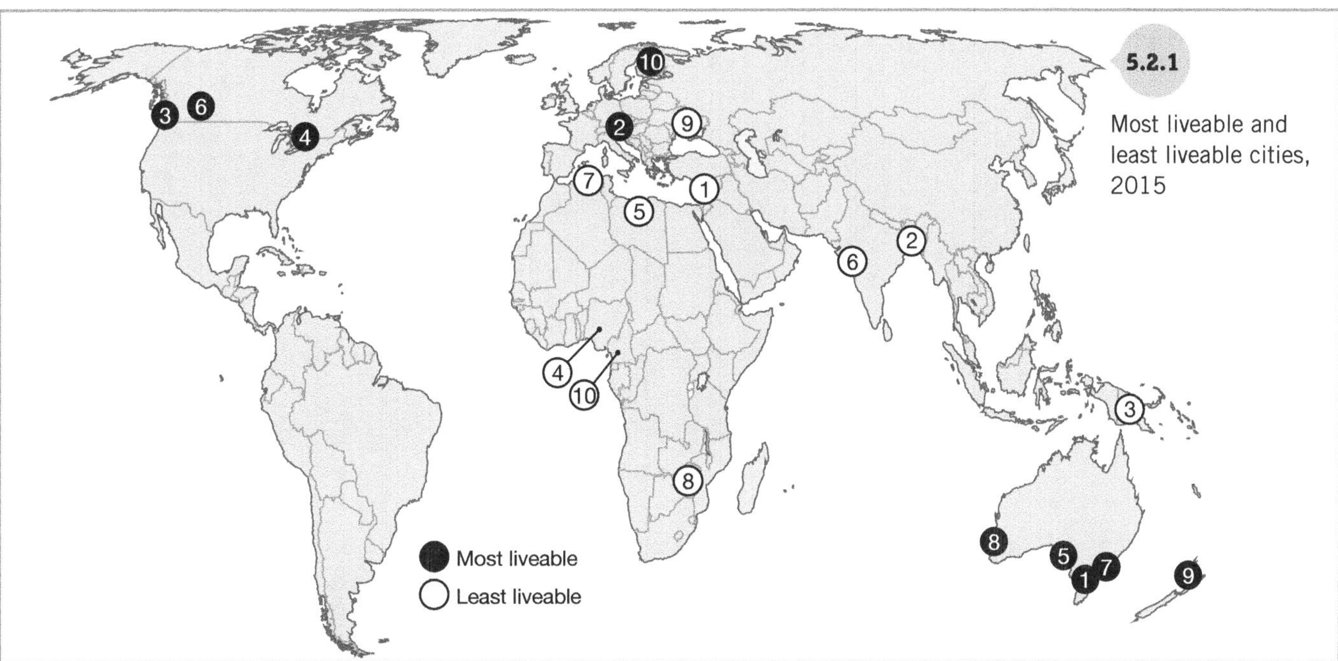

Most liveable		Rating 100 = ideal 0 = intolerable	
Rank	**Country**	**City**	**Rating**
1		Melbourne	97.5
2		Vienna	97.4
3		Vancouver	97.3
4		Toronto	97.2
5		Adelaide	96.6
6		Calgary	96.6
7		Sydney	96.1
8		Perth	95.9
9		Auckland	95.7
10		Helsinki	95.6

Least liveable		Rating 100 = ideal 0 = intolerable	
Rank	**Country**	**City**	**Rating**
10		Douala	44.0
9		Kiev	43.4
8		Harare	42.6
7		Algiers	40.9
6		Karachi	40.9
5		Tripoli	40.0
4		Lagos	39.7
3		Port Moresby	38.9
2		Dhaka	38.7
1		Damascus	29.3

1 Complete the table in Figure 5.2.1 by listing the countries that have the world's most and least liveable cities. Use an atlas to help you.

2 Describe the spatial pattern of liveability by answering the following questions.

a Identify the continents where liveability is high.

b Identify the continents where liveability is low.

3 How might the perceptions of liveability differ between people who live in rural locations and people who live in urban areas (cities)? Fill out as many differences as you can in the table below.

Rural	Urban

4 Use your answer to question 3 to construct criteria that could be used to measure the liveability in rural areas.

5 Use your answer to question 3 to construct criteria that could be used to measure the liveability in urban areas.

5.3 Crime and liveability

Knowledge and understanding

visual–spatial • logical–mathematical

Table 5.3.1 shows the number of robberies committed in March 2012 in various boroughs (city subdivisions) of London, United Kingdom. Study the table and complete the questions that follow.

Borough	Number of robberies
Barking and Dagenham	11
Barnet	14
Bexley	2
Brent	20
Bromley	24
Camden	14
Croydon	29
Ealing	25
Enfield	11
Greenwich	14
Hackney	17

Borough	Number of robberies
Hammersmith and Fulham	18
Haringey	17
Harrow	17
Havering	7
Heathrow Airport	0
Hillingdon	25
Hounslow	5
Islington	11
Kensington and Chelsea	7
Kingston upon Thames	2
Lambeth	34

Borough	Number of robberies
Lewisham	28
Merton	4
Newham	39
Redbridge	20
Richmond upon Thames	2
Southwark	38
Sutton	5
Tower Hamlets	21
Waltham Forest	22
Wandsworth	9
Westminster	18

Source: Metropolitan Police, UK

5.3.1 Number of robberies in London in March 2012, by borough

1 Use different shades of colour to represent each range in the legend below. For example, you could use white, light blue, baby blue, medium blue, navy and black. Then use the legend to shade the boroughs on the map with the correct colour.

5.3.2 Number of robberies in London boroughs

2 Describe the spatial pattern of robberies in London during March 2012.

___

___

___

3 Identify the areas in London that appear to have a higher liveability in terms of robberies.

___

___

4 Some governments publish guidelines about how to design neighbourhoods in a way that reduces crime. Study the following two pairs of illustrations. Identify the differences between them. Explain how the 'ticked' version could help reduce crime.

Landscaping and surveillance	Difference	How does it reduce crime?
 ✗		
 ✓		

Urban structure	Difference	How does it reduce crime?
 ✗		
 ✓		

5 One way to reduce crime can be to put CCTV (closed-circuit television) cameras on city streets to record activity. How might having many cameras on streets affect liveability? Identify the positive and negative impacts of CCTV on liveability by filling out the table below.

Positive impacts	Negative impacts

5.4 Definitions: Liveability

Knowledge and understanding

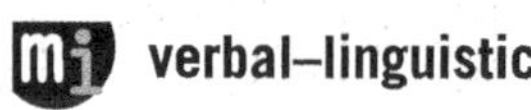

Complete each definition using the words after each sentence. Then, write the term that matches the definition, choosing from the following: *aesthetics*, *anti-social*, *quality of life*, *liveability*, *perceptions*, *spatial* and *telework*.

Term	Definition
	The ____________ of a place (city, town, suburb or neighbourhood) that contribute to the quality of ____________ experienced by those who live or ____________ there visit qualities life
	How something is ____________ or ____________ by ____________ someone regarded seen
	Any ____________ that relates to a ____________ and/or an ____________ location area characteristic
	____________ that involves the use of ____________ technologies as a substitute for ____________ travel telecommunications work physical
	A ____________ of ____________ for what a person finds ____________ or appealing criteria attractive set
	The ____________, well-being and satisfaction of a ____________. Among the many factors that influence quality of life are the person's ____________ income and access to ____________ family person services happiness
	Behaviour that lacks ____________ for others and may cause damage to the ____________ whether intentionally or through ____________. Behaviour is labeled anti-social when it is deemed ____________ to prevailing ____________ for social conduct. norms neglect society consideration contrary

5.5 Deciding where to live

Knowledge and understanding • Geographical skills

mi verbal–linguistic • logical–mathematical

Study Figures 5.5.1 and 5.5.2, which show the reasons given by people in Western Australia and Queensland for choosing to live at a particular location.

Reason	Total number of people ('000)	Total number of people (%)
Quiet location	397.4	42.5
Close to family or friends	392.3	42.0
Familiarity with area	388.0	41.5
Access to facilities and services (e.g. shops, schools)	379.9	40.6
Safe neighbourhood	377.5	40.4
Central location	363.2	38.8
Better lifestyle	352.4	37.7
Geographical features (e.g. beach, hills, river)	305.1	32.6
Close to work	291.3	31.2
Close-knit community	177.7	19.0
Limited availability of dwellings	13.9	1.5

Source: Australian Bureau of Statistics

5.5.1 Reasons for choice of current dwelling location, Western Australia

Reason	Total number of people ('000)	Total number of people (%)
Access to facilities (e.g. shops, schools)	162.0	47.0
To live close to work or to improve employment prospects	150.6	33.9
Better lifestyle	105.6	23.8
Close to family or friends	113.8	25.6
Wanted a bigger or smaller home	51.5	11.6
Bought this dwelling as a first home buyer	46.7	10.5
To reduce rental/mortgage costs	34.1	7.7
Change in family circumstances	21.8	4.9

Source: Australian Bureau of Statistics

5.5.2 Reasons for choice of current dwelling location, Queensland

5.5

1 a For Western Australia, identify the four most significant reasons people give for their choice of where to live.

____

____

____

b For Queensland, identify the four most significant reasons people give for their choice of where to live.

____

____

____

2 a Compare the top three reasons given for choosing where to live in Western Australia with the top three reasons given for Queensland.

____

____

____

____

b Suggest a possible explanation for differences between the reasons.

____

____

____

3 Read the quotes below. From the word box, identify the factor that influenced each person. Write the factors in the spaces provided.

environmental factors	personal safety	transport options
lifestyle considerations	employment opportunities	close to family and friends
affordability		

We live in this suburb because our friends live around the area. My in-laws are also only a few minutes away, which is good because they can look after our twins at fairly short notice.

____

Basically, I moved here for a job. There was work in the mine and it pays well, so I applied and got the job. The company paid moving costs, pays my rent and even pays for me to fly back to my home city a couple of times a year.

____

Of course we'd like to live in a large house near the beaches, but that is just out of our price range. So this suburb is really what is within our budget.

____

I couldn't live anywhere else. You have cafes, bars, shops and a cinema just a street away. I can go out and get home in just 10 minutes.

____

Our choice to live in this area basically came because of the crime rates where we used to live. I looked at the statistics for crime rates of other suburbs and this had the lowest one available.

____

We originally came from South-East Asia, so we looked for a place in Australia that would have a similar climate to what we are used to.

____

My wife and I made a conscious decision not to buy a car a few years ago, but it meant we had to move to here to be closer to public transport.

____

5.6 Change in Australian communities

Geographical skills

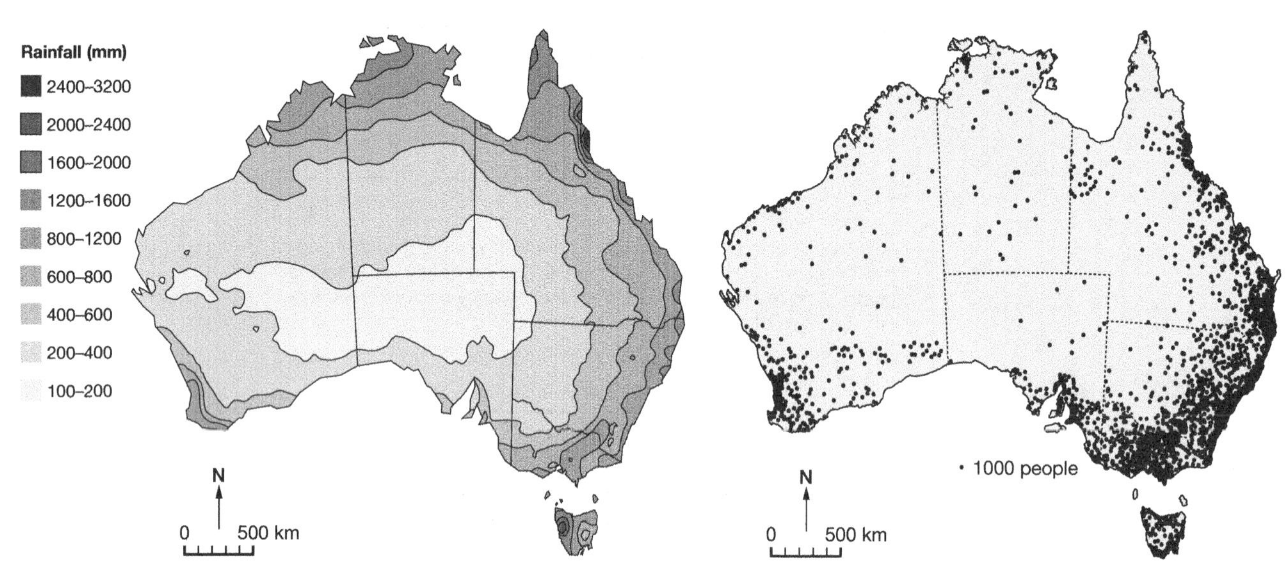

Source: Bureau of Meteorology

5.6.1 Annual median rainfall for Australia (millimetres)

Source: Australian Bureau of Statistics

5.6.2 The distribution of the Australian population

1 a Study Figures 5.6.1 and 5.6.2 and compare the distribution of Australia's population with the pattern of annual mean rainfall.

____

____

____

____

b Describe the similarities between the distribution of population and the pattern of median rainfall in Australia.

____

____

____

c Suggest why there would be a high correlation between population distribution and rainfall.

____

____

2 Compare the distribution of Aboriginal and Torres Strait Islander People with non-Indigenous Australians by referring to the graph in Figure 5.6.3.

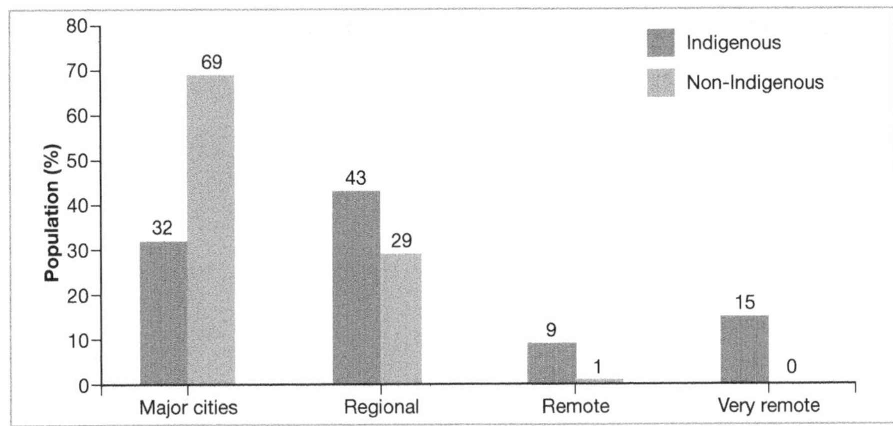

5.6.3

Population distribution in Australia by remoteness, area and Indigenous status, 2016

Source: Australian Bureau of Statistics

---

3 a Use Figure 5.6.4 to identify the total population of Australia in 2005.

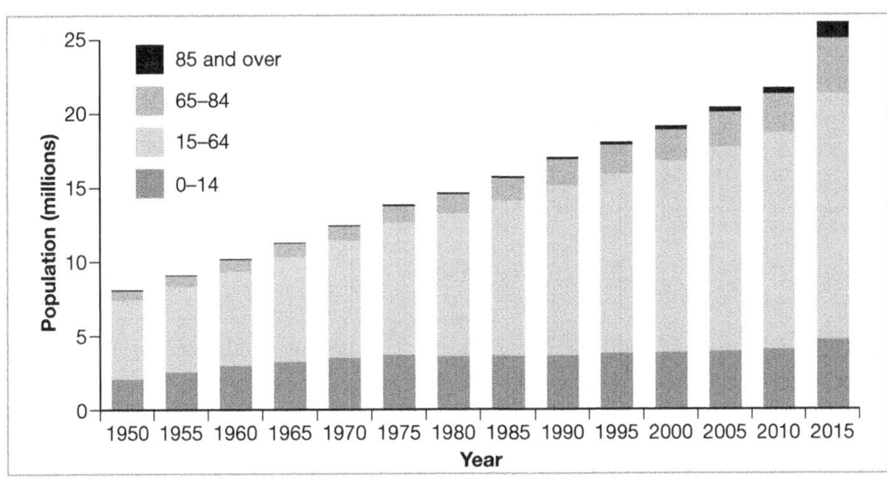

5.6.4

The changing population of Australia

Source: Australian Bureau of Statistics

b Describe at least three trends in Australia's population according to Figure 5.6.4.

---

c Which age group had the greatest percentage increase between 1950 and 2010? To calculate the percentage increase for each age group, subtract the 1950 population from the 2010 population, divide by the 1950 population and multiply by 100.

5.7 Australia's population

Geographical skills • Knowledge and understanding

verbal–linguistic • logical–mathematical

Study Figure 5.7.1 and answer the questions that follow.

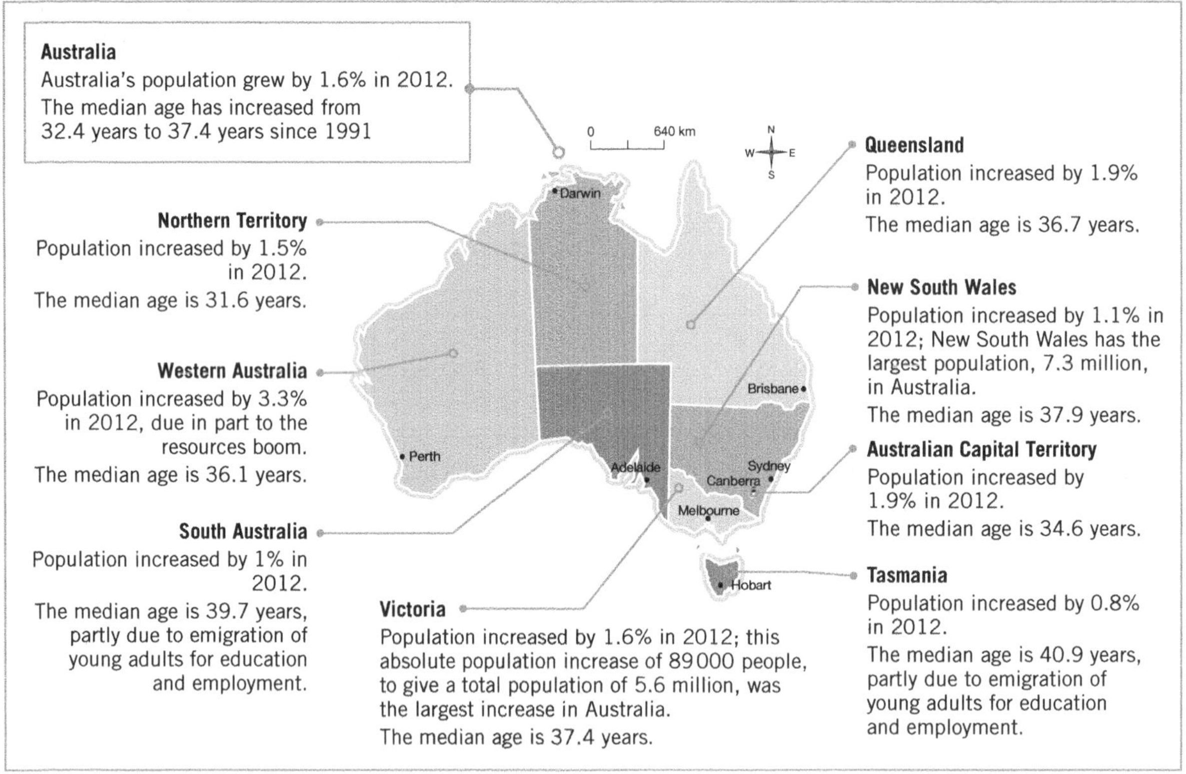

Source: Australian Bureau of Statistics

5.7.1 Australia–one of the world's fastest growing nations

1 a The median age is the age at which half the population is younger and half is older. Compare the median age of Tasmania and South Australia with that of Australia as a whole.

____

____

b Provide one reason suggested by the Australian Bureau of Statistics (ABS) for the median ages in Tasmania and South Australia.

____

____

c Identify the two states with the lowest median ages.

____

d Identify the population growth rates for Victoria, Queensland and Western Australia.

____

____

e Identify the state with the highest population growth rate and the state with the greatest increase in population.

____

____

f Identify the reason for population growth in the fastest-growing Australian state.

____

____

g Propose two possible impacts of an ageing population for South Australia.

____

____

____

Refer to the graph in Figure 5.7.2 to answer question 2.

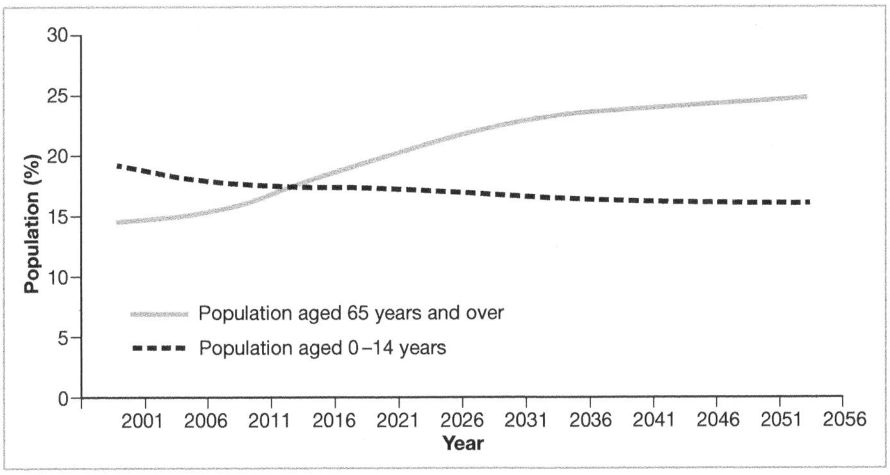

Source: Australian Local Government Association, 'Australia's ageing population: Economic implications for local government', 2004

5.7.2

Projections of the proportion of children aged 0–14 years and people aged over 65 in South Australia, from 1998 to 2056

2 a Identify the current proportion of the population aged between 0 and 14 years in South Australia.

____

b According to the projections of the graph, in what year will the number of people between the ages of 0 and 14 years overtake the number of people aged over 65?

____

c Describe the population trend for South Australia.

____

____

____

5.8 Population distribution – the coast

Knowledge and understanding

verbal–linguistic

Read the text below and answer the questions that follow.

'Sea change' refers to the shift of people from urban areas to coastal communities. The National Sea Change Taskforce (NSCT) has been set up to represent coastal communities that have experienced rapid growth and development resulting from people migrating to the coast.

According to the NSCT, almost 6 million people live in Australian coastal areas outside the capital cities. The growth rate in these coastal areas is over 60 per cent higher than the national average. NSCT is concerned that such a high growth rate can't be sustained.

Local councils in these areas do not have the resources to keep pace with the growth. The growth will mean an increasing demand for infrastructure such as roads, water and sewerage. Services such as public transport, health care and education are also inadequate and expensive to expand.

According to the head of the NSCT, baby boomers (people born between 1946 and 1964), make up the largest demographic group in Australia. Many baby boomers have now reached retirement age. About one million baby boomers plan to move to the coast over the 16 years following 2010.

However, it is not only retirees who are undergoing a sea change. The data in figure Figure 5.8.1 suggest that many other groups of Australians are moving to the coast.

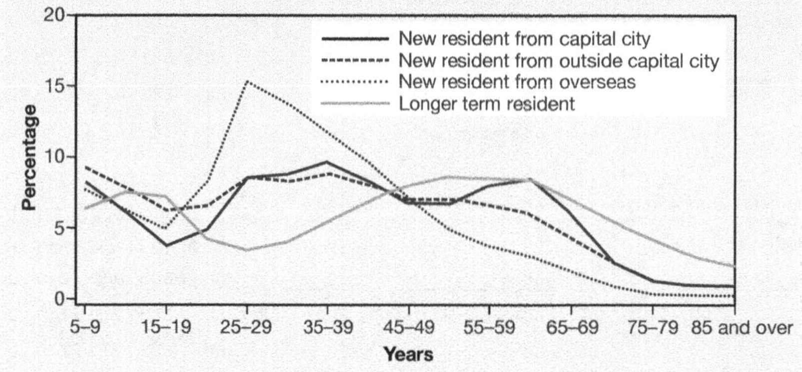

5.8.1 Age profile of residents of country coastal areas

Source: Australian Bureau of Statistics, 2013

1 Explain the term 'sea change'.

____

____

2 Study Figure 5.8.1 and answer the following questions about residents of country coastal areas in Australia.

a Identify the proportion of residents aged 5–9 years who were from a capital city. ____________

b Identify the proportion of residents aged 65–69 years who were from a capital city. ____________

c Identify the proportion of residents aged 65–69 years who are longer-term residents. ____________

3 Discuss whether this statement is correct: 'Most people who are moving to the coast are retired people looking for a sea change.'

4 What are the challenges facing coastal communities as their population increases?

5 Some factors that people usually consider when choosing where to live are climate, landscape, recreation and community. In the diagram below, describe what makes coastal areas so appealing in terms of these factors.

Climate	Landscape
Recreation	Community

5.9 Population distribution – economic activity

Knowledge and understanding

verbal–linguistic • visual–spatial

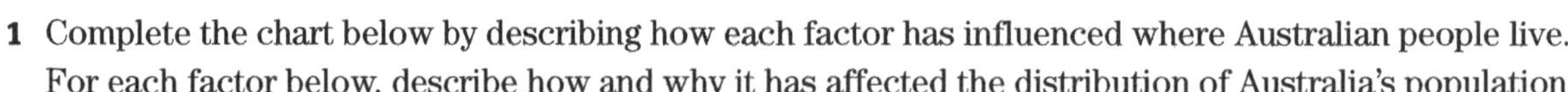

1 Complete the chart below by describing how each factor has influenced where Australian people live. For each factor below, describe how and why it has affected the distribution of Australia's population.

Factor	Impact on population distribution
Financial institutions	
Pattern of land use	
New technologies and the global economy	
Trade	
Mining boom	

6.1 Enhancing liveability

Knowledge and understanding

 verbal–linguistic • intrapersonal

1 Explain how the following factors might affect someone's perception of liveability.

a Age ______________________________

____

b Income ______________________________

____

2 Suggest how the concept of liveability might be different for people who live in cities and people who live in country areas.

____

____

____

3 Think about the city where you live, or the city nearest to where you live. How could the liveability of this city be improved? List some ways in which you think liveability could be improved in each of the categories in the diagram.

Lifestyle	**Community**	**Open space**
Streetscape		**Transport**
Accessibility	**Natural environment**	**Housing**

4 The city in Figure 6.1.1 below is Chicago, United States. The city council has developed a plan to create a more liveable city by the year 2040. The table below outlines some strategies that the council is using to try to achieve this. Complete the table by explaining how each of the strategies will improve liveability.

Strategy	How will it improve liveability?
Develop more parks and open spaces	
Increase availability of sporting facilities	
Increase energy efficiency of buildings and cars	
Develop cycle paths, pavements and public transport	
Redevelop existing communities rather than creating new suburbs a long way from the city	

5 Add labels to the photograph in Figure 6.1.1 that indicate how the liveability of this area of Chicago could be improved.

6.1.1 Chicago, United States

6.2 Better recreational spaces

Knowledge and understanding

 verbal–linguistic • visual–spatial

1 Discuss how better recreational spaces can improve the liveability of an area.

2 Opportunities to create better recreational spaces often arise when land is vacant. Study the photograph in Figure 6.2.1. Imagine you are in charge of developing a recreational space on this vacant land. Write a questionnaire that is designed to find out the needs and characteristics of the community, so that the new recreation area will best improve liveability. Think carefully about what information you want to find out and what questions you need to ask to get the information.

6.2.1

Inner city vacant block. This land is to be developed into a recreational space for the local community

6.2

Read the community profile below then answer the following questions.

Population size	6342
Percentage born overseas	18%

Age	
0–4 years	10%
5–12 years	20%
13–18 years	15%
19–30 years	12%
31–50 years	27%
51–64 years	12%
65+ years	4%

Family	
Single parent only	15%
Both parents living together	47%
Single parent with extended family	12%
Both parents with extended family	5%
Unmarried and unrelated family	15%
Other	6%

3 **a** Design a recreational space for the vacant area that will best meet the community's needs. The area available measures 100 metres square. Draw a plan for your design in the space provided. Remember to include BOLTSS and label key features.

b Explain how the key features of your design will improve liveability for the community residents. In your answer, explain how your design considered the age structure and the family type of the community.

7.1 Weather and climate

Knowledge and understanding • **Geographical skills**

1 Describe the difference between 'weather' and 'climate'. Give an example to support each description.

2 Describe the climate for this time of year for where you live.

3 Below are statements about weather and climate. Identify which refer to weather (W) and which describe climate (C).

- [ ] It is so hot outside this afternoon
- [ ] Don't forget your umbrella. It's going to rain today.
- [ ] It's hot, steamy and wet in Bali.
- [ ] Tourists visit Hawaii all year round to enjoy the sunny beaches.
- [ ] It is always very cold in the North Pole.
- [ ] The wind may change direction tomorrow and bring cold and rain.
- [ ] In New Zealand, people go skiing during winter.

4 Define atmospheric pressure.

5 Describe what happens to the atmospheric pressure over a particular place when air is heated.

6 Knowing the atmospheric pressure over a particular place allows us to make predictions about what the weather will be there. Describe the predicted weather for areas of high and low pressure.

a Areas of high pressure

b Areas of low pressure

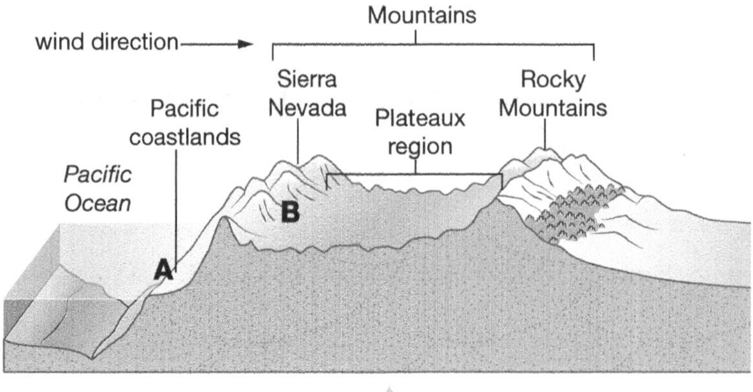

7.1.1 Cross-section of North America

7 **a** Use your knowledge of weather to describe the likely temperature and rainfall at locations A and B in Figure 7.1.1.

A ______________________________

B ______________________________

b Explain the differences in temperature and rainfall between locations A and B.

____

____

8 In pairs, discuss how the weather conditions listed in the table below may affect people living in Australia. Then write your answers in the table.

Weather condition	Example of positive impact	Example of negative impact
Heavy rain		
Drought		
Windy weather		

7.2 The water cycle

Knowledge and understanding

verbal–linguistic

1 Use words from the box below to complete the following paragraphs.

hydroelectricity	dew	circulating	agriculture	form	sun	gravity
groundwater	location	atmosphere	liquid	streams	fog	gas
precipitation	bores	infiltration	unreliable	surface	land	rivers
condensation	dams	84 per cent	closed			

Movement and change

The same water has been ____________________ on the earth since the earliest days of the planet, driven by ____________________ and the energy of the ____________________. No water is added or removed from the water cycle and so it is called a ____________________ system. Water is always moving and changing its ____________________ and geographical ____________________.

Water in the atmosphere

Water changes from a ____________________ to a ____________________ in a process known as evaporation. About ____________________ of the water vapour in the ____________________ comes from the oceans. The process of ____________________ occurs when the water vapour changes into liquid water in the form of ____________________, ____________________ or cloud droplets.

Water returns to the earth

When the water falls to the ground it is called ____________________ and may be either solid or liquid. Only 17 per cent of all precipitation falls on the ____________________. Once on the land, water may move into the soil, known as ____________________, or flow on the ____________________ of the land into ____________________, ____________________ and lakes, which is known as run-off.

Water storage and human use

Water stored beneath the earth's surface is known as ____________________ and is an important source of water for ____________________, through the use of wells and ____________________.

Other ways humans interact with the water cycle include storing water in ____________________ where rainfall is ____________________; as well as using water and gravity to generate ____________________.

7.3 Precipitation

Knowledge and understanding

 verbal–linguistic • visual–spatial

1 In the diagram, write what happens in order for each type of precipitation to occur.

Read the article about condensation and the water cycle below, then answer the questions that follow.

The water cycle: condensation

Condensation is the process by which water vapour in the air is changed into liquid water. Condensation is crucial to the water cycle because it is responsible for the formation of clouds. These clouds may produce precipitation, which is the primary route for water to return to the earth's surface within the water cycle. Condensation is the opposite of evaporation.

Condensation occurs when warm moist air comes into contact with a cold surface.

Clouds exist in the atmosphere because of rising air. As air rises and cools the water in it condenses, forming clouds. Since clouds drift over the landscape, they are one of the ways that water moves geographically around the globe in the water cycle.

7.3.1 Condensation on a cold drink bottle

2 In your own words, define condensation.

____

____

3 How does the process of condensation explain the presence of water on the outside of a cold, unopened drink bottle like the one in Figure 7.3.1?

____

____

____

4 The following diagram shows three ways that rain can form. Complete the table by identifying the type of rainfall being illustrated, the main factor in forming the rain, and by describing how the rain is formed.

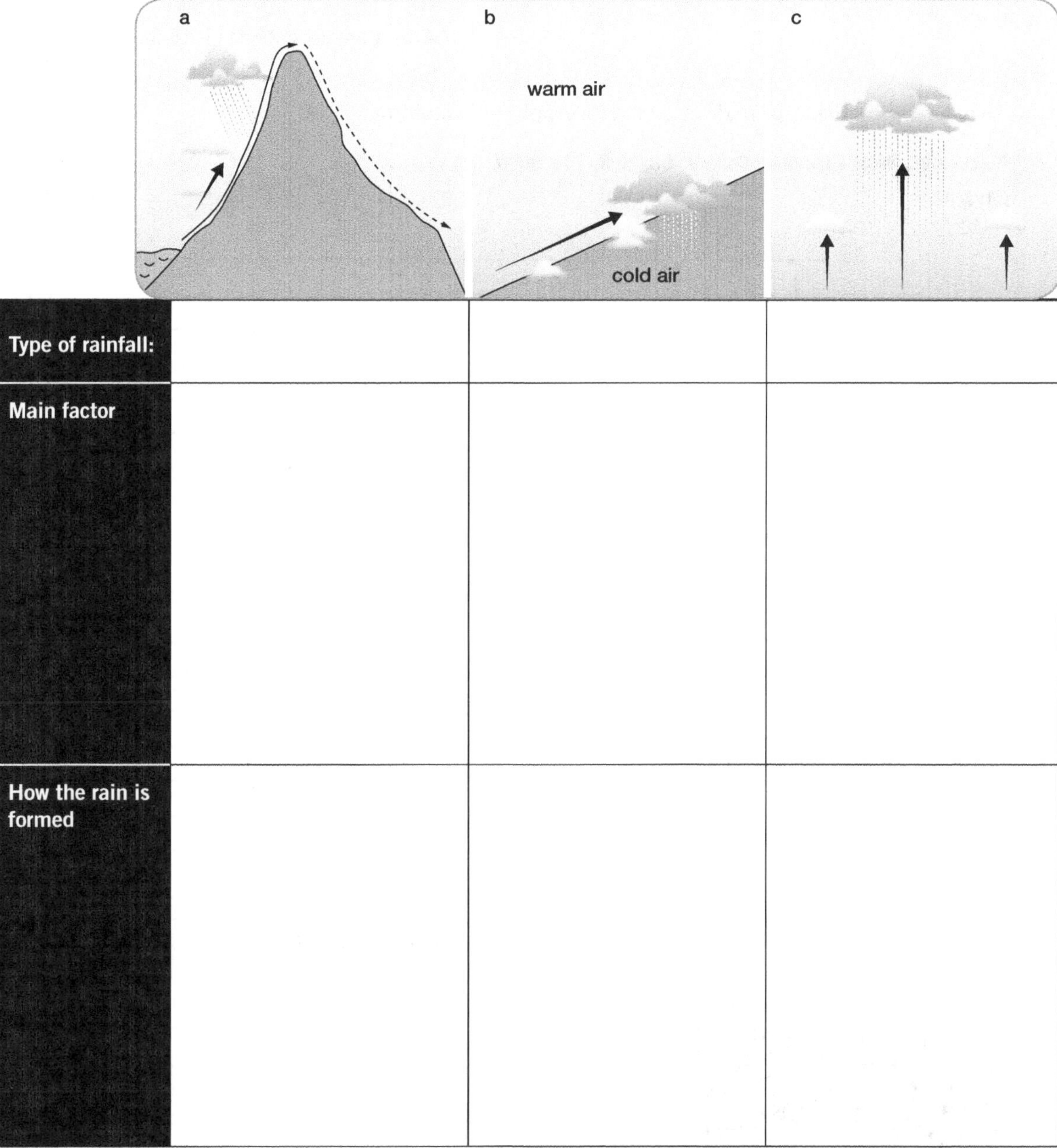

	a	b	c
Type of rainfall:			
Main factor			
How the rain is formed			

7.4 Geoskills: Climate graphs

Geographical skills

mi **logical–mathematical • visual–spatial**

Refer to Table 7.4.1 which shows climate statistics for Edmonton, Canada to complete the questions that follow.

	J	F	M	A	M	J	J	A	S	O	N	D
Average temperature (°C)	-11.7	-8.4	-2.6	5.5	11.7	15.5	17.5	16.6	11.3	5.6	-4.1	-9.6
Precipitation (mm)	22.5	14.6	16.6	26	49	87.1	91.7	69	43.7	17.9	17.9	20.9

7.4.1 Climate statistics for Edmonton, Canada

1 a Use the data in the table to draw a climate graph for Edmonton.

Climate graph for Edmonton, Canada, 53°32'N, 113°30'W

TEMP °C	PRECIPITATION mm
24	95
22	90
20	85
18	80
16	75
14	70
12	65
10	60
8	55
6	50
4	45
2	40
0	35
-2	30
-4	25
-6	20
-8	15
-10	10
-12	5
-14	0

Jan Feb Mar Apr May Jun Jul Aug Sep Oct Nov Dec

b Describe the climate of Edmonton by filling out the table below. You will need to use a calculator.

Annual temperature range	
Average annual temperature	
Annual precipitation	
Average annual precipitation	

7.5 Climates around the world

Geographical skills

 visual–spatial

1 Below is a world map. Four countries – Kenya, Greenland, the United Kingdom and New Zealand – have been highlighted. For each country, there is a box containing four types of information:

A – location of the country

B – type of climate

C – general temperature pattern throughout the year

D – rainfall pattern throughout the year

Visit a website such as www.climate-zone.com to complete the information for Greenland, the United Kingdom and New Zealand. Kenya has been completed for you.

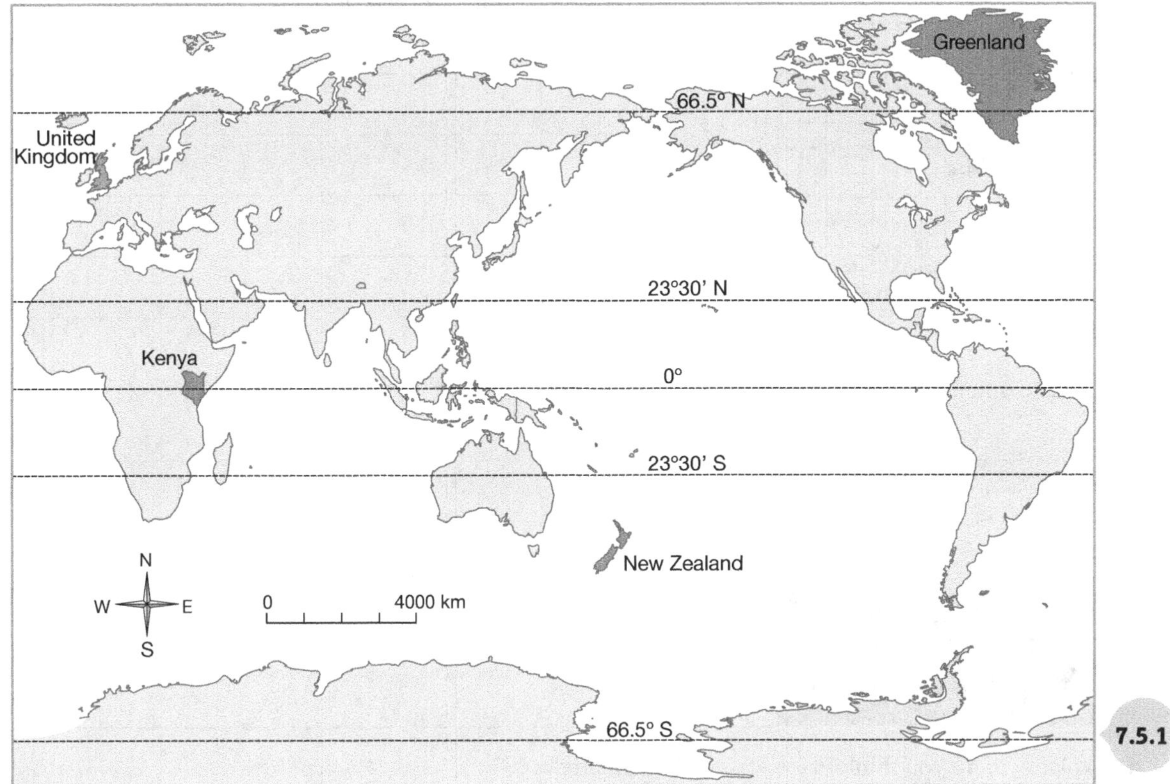

7.5.1

Kenya

A	On the Equator around 0°
B	Tropical climate
C	High all year round
D	High annual rainfall

United Kingdom

A	
B	
C	
D	

Greenland

A	
B	
C	
D	

New Zealand

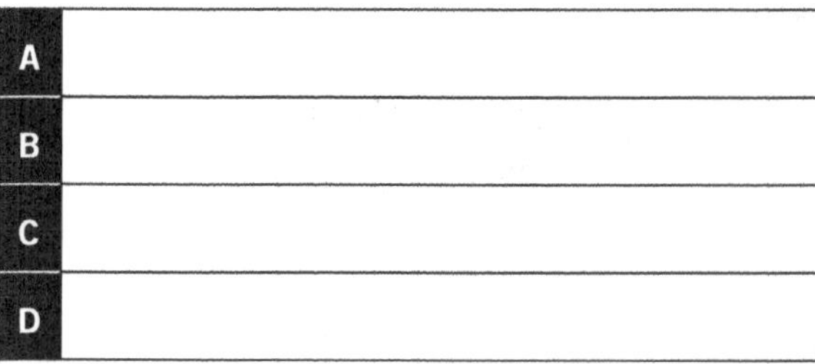

A	
B	
C	
D	

2 Below is a climate graph of Place X in a particular year. Guess the location of Place X and the type of climate it experiences. Answer the following questions to help you.

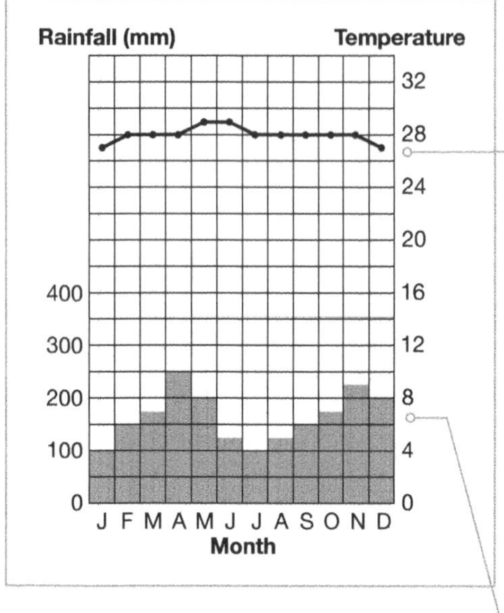

a What do you notice about the temperature?

____

____

____

____

____

b What do you notice about the rainfall?

____

____

c Complete the following information:

Location of place (in latitude)	
Type of climate	

d Explain where you think Place X would be located.

____

____

7.6 Geoskills: Weather maps

Knowledge and understanding

 **verbal–linguistic • visual–spatial**

Study the weather map in Figure 7.6.1 and complete the questions that follow.

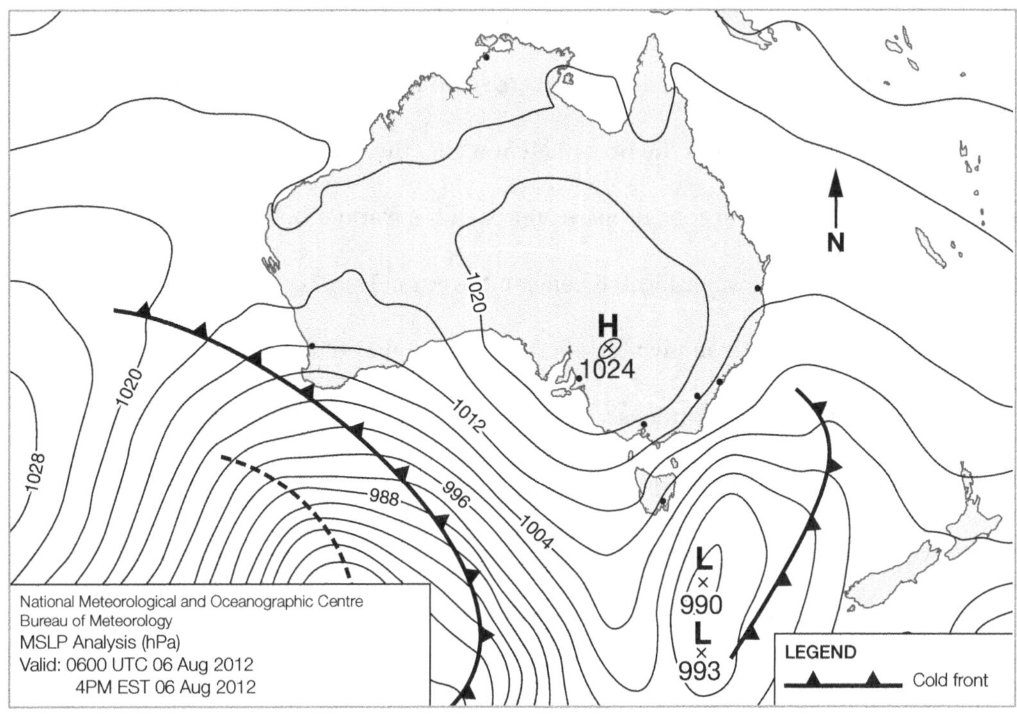

7.6.1 Weather map of Australia at 4.00 pm, 6 August 2012

1 a Name the type of fronts present on the weather map.

____

b Describe the location of each of the fronts.

____

2 Outline the weather conditions that would result from the high pressure system over Australia.

____

____

3 Identify the air pressure at Melbourne. ______________________________

4 Determine whether wind speed will be greater in Sydney or Perth.

____

5 Predict what the weather will be like in the next 24 hours in:

a Perth ______________________________

b Sydney ______________________________

7.7 Definitions: Weather and climate

Knowledge and understanding

 **verbal–linguistic**

1 Circle the correct terms in the sentences below.

Today, the meteorologist/cartographer studied the processes in the earth's atmosphere. She suggested that the mountain range near the central coast would receive orographic/frontal as warm, moist air approaches from the ocean. Meanwhile, the southern seas could get orographic/frontal rainfall as the cool air mass approaches a warmer body of air. Hopefully, this will increase the amount of water located beneath the ground surface in soil pore spaces, a feature known as rain shadow/groundwater . In the northern part of the country, the amount of water vapour in the air, known as condensation/humidity , was much higher than in previous weeks. Today's weather/climate was quite different from yesterday's. She predicted that a storm cell/rain shadow would have updrafts and downdrafts in convection loops. The rapid supply of groundwater/precipitation , mainly in the form of heavy rain, would lead to higher-than-normal amounts of atmospheric pressure/run-off , as water travels down the slope.

2 Create a concept map linking the following key terms. You can add others terms too.

weather • climate • rainfall • precipitation • temperature • humidity • wind

7.8 Dams and catchments

Geographical understanding • Case study

 **verbal-linguistic • visual spatial**

The Hoover Dam

Why?
- The United States needed to control floods, provide irrigation water for the surrounding arid plains and produce hydroelectric power.

How?
- Before construction could occur, the water of the Colorado River needed to be diverted. Four 17-metre-wide tunnels were constructed in the canyon walls to divert water around the construction site.
- If the concrete for the dam was poured all at once, it would have taken 128 years to dry. As a result, concrete was poured as a series of many smaller blocks, each with pipes of running cold water within them to dry the concrete quickly.

What?
- The Hoover Dam is one of the world's biggest concrete dams.
- It is 221 metres high and weighs an estimated 6.6 million tons
- Lake Mead was created by the Hoover Dam. This human-made lake, which is 180 kilometres long with a shoreline of 890 kilometres, is one of the longest in the world.

Positive effects of the dam
- Controls floods
- Provides irrigation for farmland
- Provides domestic and industrial water for over 14 million people
- Clearing of previously muddy water in reservoirs and parts of the Colorado River
- Provides recreation for over 10 million visitors per year to the lake
- Provides new habitats for fish and wildlife
- Generates low-cost, renewable, clean hydroelectric power for use in Nevada, Arizona and California
- Produces about 4 billion kilowatt-hours of energy is per year, enough for 500 000 homes

Where?
- The Hoover Dam is located in the Black Canyon of the Colorado River in the United States, on the border between the states of Arizona and Nevada.

7.8.1 Hoover Dam, United States

When?
- Planning for the dam began in 1922, construction began in 1931 and the last concrete was poured in 1935.
- The first hydroelectric generator began operation in October 1936, while the seventeenth and last generator began in 1961.

Negative effects of the dam
- Caused increases in the temperature, salinity and levels of chemical pollution of the Colorado River
- Contributed to the decline of estuary ecosystems like the Colorado River Delta where the river enters the Gulf of California. Wetlands here are about 5 per cent of their original extent because of the reduced flow of water from the river.
- Species of plants and animals (especially fish) that rely on the regular natural flooding have become endangered.

1 **a** Different groups have different opinions about the construction and operation of the Hoover Dam in Figure 7.8.1. Choose two of the following groups and outline what their opinion of the dam might be: industrialists/factory owners; environmentalists; farmers; residents of Arizona and Nevada.

Group 1:

____

____

Group 2:

____

____

b Discuss the impact of the dam on the operation of the water cycle at this location.

Catchment

2 Study the following diagram of a catchment.

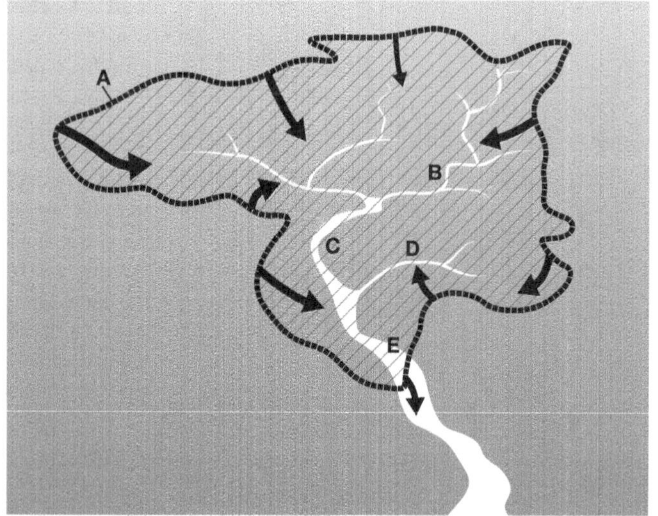

a Identify the feature labelled A.

b Identify the area that is shaded with diagonal lines.

3 Describe the drainage pattern of this river system.

4 Water was tested at C and E. The water at C was not polluted, while the water at E was heavily polluted.

a Is the most likely source of pollution tributary B or D? ______

b Justify your answer. ______

7.9 Water footprint

Geographical skills

 verbal–linguistic • logical–mathematical

1 Refer to Figure 7.9.1 to answer the following questions.

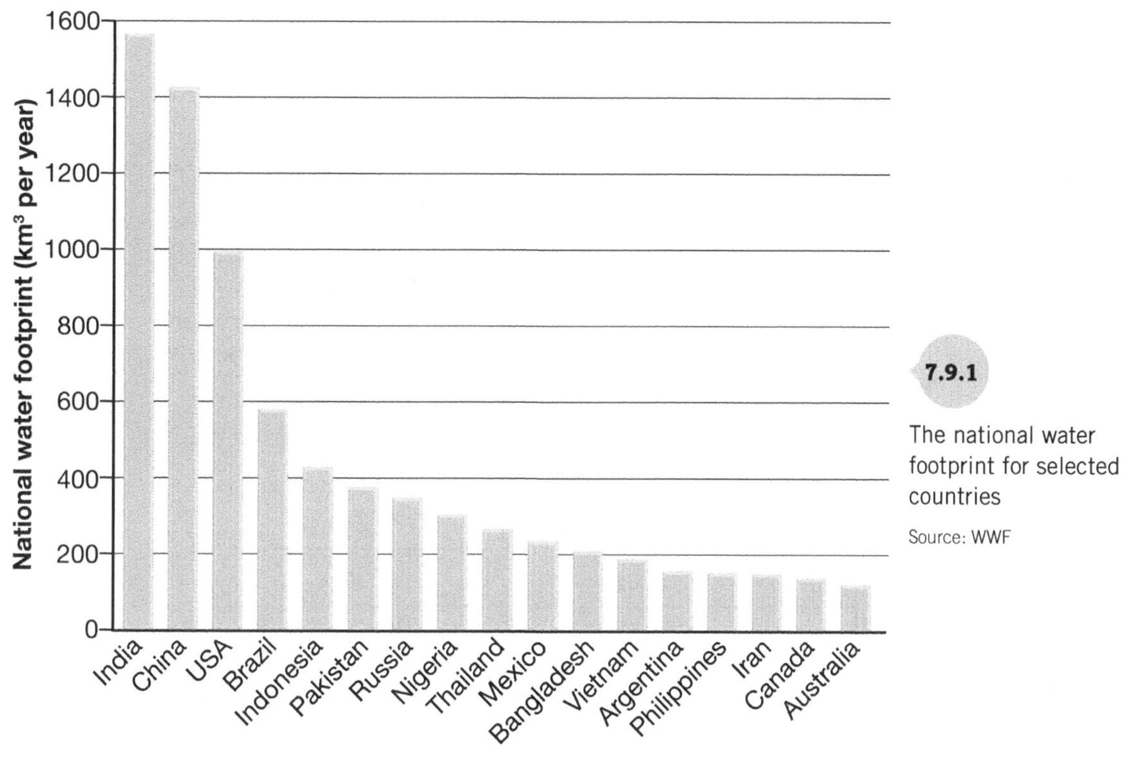

7.9.1

The national water footprint for selected countries

Source: WWF

a Identify which country has the highest national water footprint.

____

b Propose why this country has the highest national water footprint.

____

____

____

2 The following table shows the water footprint per capita (the average water footprint for each person in the population) for the countries in Figure 7.9.1. Note that 1 km^3 = 1 000 000 000 m^3.

Country	Population (millions)	National water footprint (km^3 per year)	Water footprint per capita (m^3)
India	1189	1564	1315
China	1338	1428	1067
USA	310		3219
Brazil	193	584	3026
Indonesia	235	431	1834
Pakistan	185	384	2075
Russia	142	355	
Nigeria	158	309	1956
Thailand	68	268	3941
Mexico	111	238	2144
Bangladesh	164	209	1274
Vietnam	89	192	2157
Argentina	41	162	3951
Philippines	94	158	1681
Iran	75		2053
Canada	34	144	4235
Australia	22	119	

7.9.2

Source: Population Reference Bureau, World Population Data Sheet

3 Use Figure 7.9.1 to fill in the missing data for the United States of America and Iran.

4 Complete the column for water footprint per capita for Russia and Australia. To do this, multiply the national water footprint figure by 1000, and then divide by the population in millions.

5 Do the countries with the highest national water footprint have the highest per capita water footprint? Propose reasons for your answer.

6 Suggest why geographers would find the water footprint per capita figures for each country useful.

7 How might educating people about their water footprint help address water shortage?

7.10 Case study: Mekong Delta

Case study

 **verbal–linguistic • visual–spatial**

Read the information below and answer the questions that follow.

The Mekong Delta is where the Mekong River empties into the South China Sea in Vietnam. It covers an area of 39 000 km^2, much of which is covered by water. The area is a valuable store of fresh water and has a long history of human use. It is a large rice-growing area, producing half of Vietnam's rice output, helping to make Vietnam the second highest exporter of rice in the world.

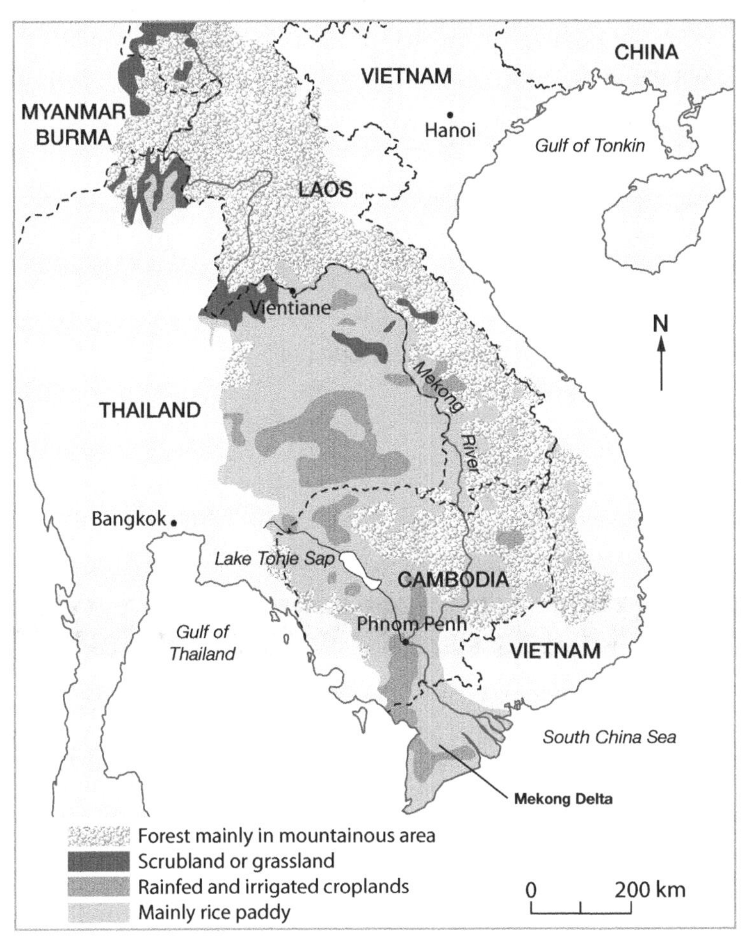

7.10.1

Location and land use map of the Mekong River and Mekong Delta

Source: The Mekong River Commission

In the middle and upper parts of the delta, fresh water is available all year round, allowing farmers to grow up to three rice crops each year. Development projects between 1999 and 2007 enabled agricultural production to increase in areas near the sea which had previously been affected by salt water. The development projects included new irrigation and drainage infrastructure, flood protection and salinity control. Drinking-water wells were also constructed.

By 2007, the average yield of rice rose from 4.7 tonnes per hectare to 5.3 tonnes per hectare. Over 760 000 rural people benefitted from safe drinking water, and household income rose. For example, income for an average farm household with 1.2 hectares increased from US$631 to US$2826 per year. The area under rice paddy irrigation grew from 4.744 million hectares in 1961 to 7.305 million hectares in 2007.

In recent years, there has been a rapid increase in the population, economy and demand for electricity in the Mekong region. To meet this need, plans have been made to build a number of new dams and hydroelectric power plants. Hydropower has many advantages over other power sources, such as fossil fuel. The power generated by new hydroelectric dams could also be sold to other countries.

However, such development could also have negative impacts. In particular, the expansion could destroy the regions' inland fishing industry and have a devastating effect on the millions of people who rely on fishing to make a living and for food.

The Mekong River, home to over 800 species of fish, is the world's largest freshwater fishery. Major dam construction would prevent the migration of fish and disrupt their breeding patterns. Scientists have estimated that up to 25 per cent of the migratory fish in the Lower Mekong Basin could die. This reduction in fish numbers has the potential to reduce biodiversity and cause great damage to the ecosystem.

Over 65 million people live in the Mekong River Basin, and around two-thirds of them rely on fish as a main source of protein in their diet. The development of large hydroelectric dams in the region would threaten the ability of millions of people to obtain food.

1 Describe how water is being used in the Mekong River and Delta.

____

____

____

2 Explain why the Mekong River and Delta is considered a 'world water hotspot'.

____

____

3 Outline the advantages and disadvantages of developing hydroelectric power within the Mekong region.

Advantages	Disadvantages

4 For each advantage and disadvantage in the table, state whether the impact is environmental, economic or social.

7.11 Definitions: Water

Knowledge and understanding

 **verbal–linguistic**

1 Draw a line to match the correct term with the corresponding definition.

Term	Definition
aquifer	transferred by contaminated water
desalination	infrastructure related to the collection and disposal of sewage (human waste)
infiltration	the processes by which water circulates between the earth's oceans, atmosphere and land
irrigation	a layer of permeable rock that is capable of storing significant quantities of water
sanitation	the loss of water vapour from parts of plants
solvent	the level at which rock strata are saturated by water
transpiration	removal of salts from seawater or other saline (salty) solutions
waterborne	the movement of water from the land surface into the soil
water cycle	a substance that dissolves other substances, thus forming a solution
water table	watering of crops by artificial means

2 Complete the 'water' phrase to match the definitions below.

Water generated from domestic activities such as washing clothes, washing dishes and bathing, which can be recycled

________________ water

The volume of fresh water that is consumed (or polluted) when a product is created

________________ water

The precipitation on land that does not run off or become groundwater, but is stored in the soil or temporarily stays on top of the soil or in vegetation

________________ water

Fresh surface water and groundwater, i.e. the water in freshwater lakes, rivers and aquifers

________________ water

8.1 Australia's water resources

Knowledge and understanding • Geographical skills

verbal–linguistic • logical–mathematical

1 Use words from the box below to complete the following paragraph.

rivers	countries	infiltrates	stored	monsoon
intercepted	two-thirds	plants	rainfall	evaporates

The average annual ________________ for Australia of 469 millimetres may not be low compared to other ________________, however, most of this is not easily ________________ because it ________________, ________________ into the ground or is ________________ and transpired back into the atmosphere by ________________. Thus, just 12 per cent of the total rainfall is run-off collecting in ________________. Northern Australia has about ________________ of Australia's run-off. Unfortunately, the rainfall in Northern Australia is seasonal, with most falling during the summer ________________.

2 Explain why seasonal rainfall in northern Australia would be difficult to use as a resource.

____

____

The Stirling Dam supplies water to metropolitan Perth. Study the table and graph below and complete the questions that follow.

Stirling Dam	
Storage as at 31 August 2012	24 304 ML (66.47%)
Storage capacity	36 563 ML
Type of construction	earthfill
Year built	1948
Catchment area	251 km^2 (25 100 hectares)
Surface area	381 ha
Wall height	46 m above ground level
Crest length	274 m
Spillway type	Weir control with unlined chute
Spillway capacity	275 m^3/s

Source: Water Corporation

8.1.1 Stirling Dam details August 2012

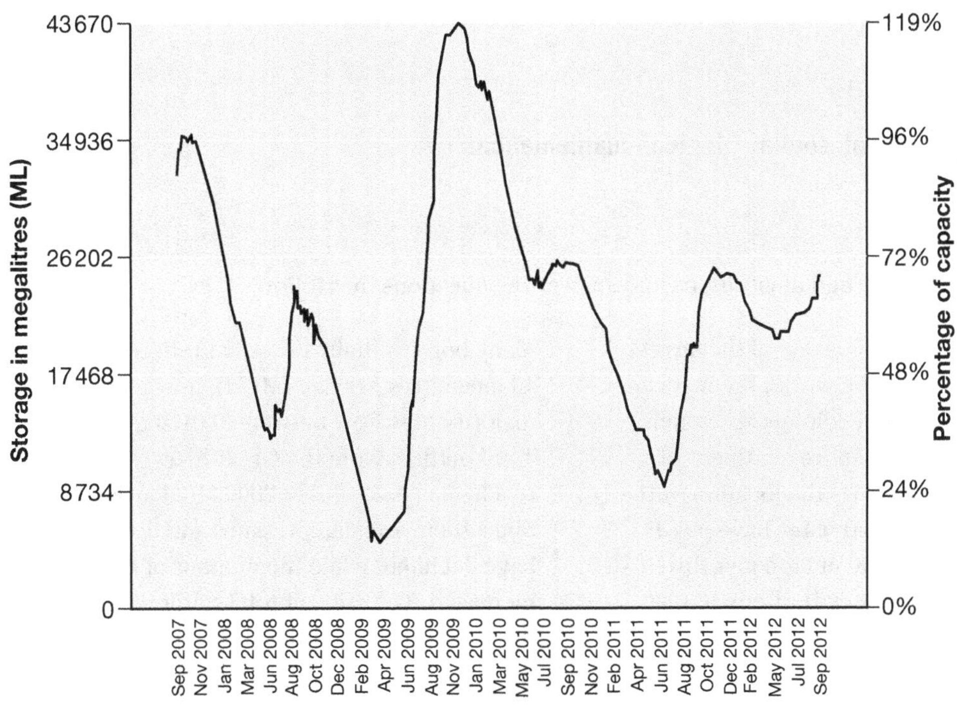

8.1.2 Storage at Stirling Dam from September 2007 to September 2012

3 What is the storage capacity of the Stirling Dam?

---

4 Use Figure 8.1.1 to find the storage of the Stirling Dam on 31 August 2012.

---

5 Suggest what caused the variations in storage amounts between 2007 and 2012.

---

6 Prepare a plan for your home or school to reduce water consumption. Provide a sketch and annotate the changes you would make.

8.2 Case study: Great Artesian Basin

Case study

 **visual–spatial • logical–mathematical**

Study the following article and illustrations and answer the questions that follow.

The Great Artesian Basin is one of the largest groundwater basins in the world. It covers an area of 1.7 million square kilometres, which is about one-fifth of the entire continent of Australia. Figure 7.3 shows the location of the Great Artesian Basin. Estimates have put its volume of water at 64 900 million megalitres—enough water to fill Sydney Harbour 130 000 times. Artesian water first entered the basin almost 2 million years ago.

The Great Artesian Basin consists of layers of sedimentary rock … Sixty million years ago, the eastern edge of the basin on the Great Dividing Range experienced uplift. Since then, rainclouds have been crossing the continent, uplifting and falling as rain on the western slopes of the Great Dividing Range. The water percolates down into the porous aquifer of the basin. Its rate of flow is estimated to be only 1–5 metres per year, gradually building up pressure under the weight of further incoming water.

Across the basin there are natural breaks … resulting in natural springs … The Cretaceous cap is also broken by human-made bores. Water rises at these points due to pressure. When pressure forces the artesian water to the surface, it is known as an artesian bore; when the water does not rise to the surface, the term 'subartesian' is applied. Artesian and subartesian bores are shown in Figure 8.2.1.

Many bores initially flowed at rates of over 10 megalitres per day (ML/d); however, the majority now flow between 0.01 and 6 ML/d. Total outflow from the Great Artesian Basin reached a peak of over 2000 ML/d around 1915. Since then, artesian pressure and flow rates have declined, while the number of bores has increased. The current total outflow from the basin is about 1500 ML/d.

Also, about one-third of all artesian bores which once flowed when drilled have now ceased to flow and require pumps to bring the water to the surface.

Source: Queensland Government

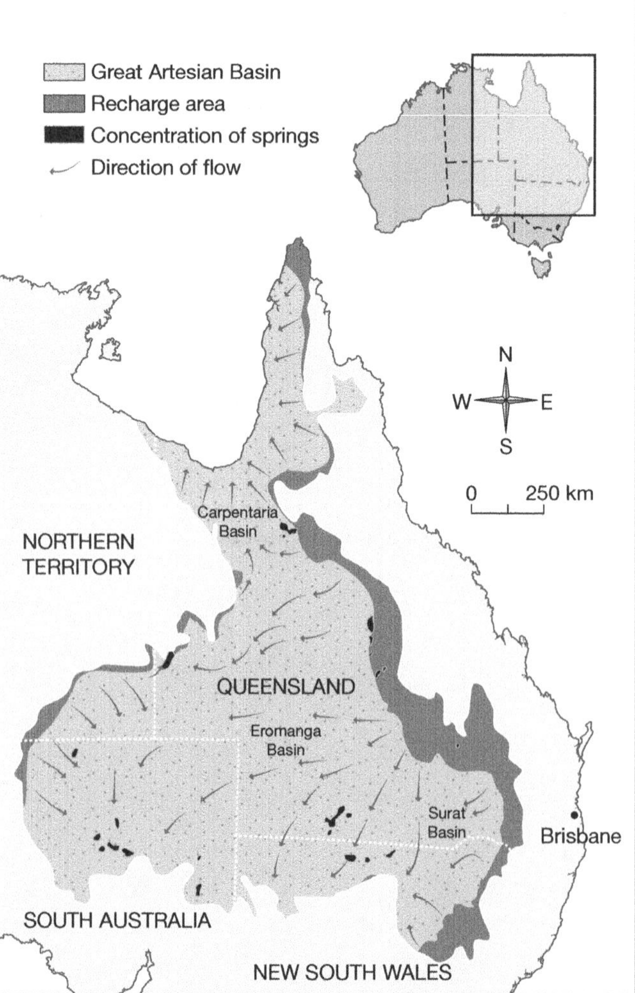

8.2.1 Location of the Great Artesian Basin

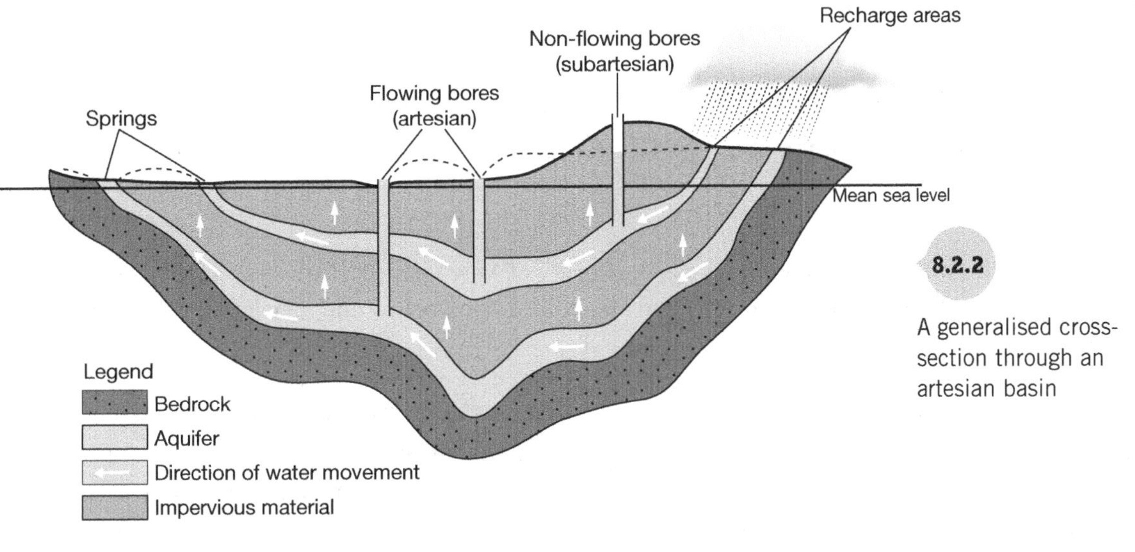

8.2.2

A generalised cross-section through an artesian basin

1 Describe the location of the Great Artesian Basin.

______

______

______

2 Explain the term 'recharge area'.

______

3 Write one sentence to describe the following features of the Great Artesian Basin.

a Size: ______

______

b Water volume: ______

______

c Age: ______

______

d Source: ______

______

e Geology: ______

______

f Natural springs: ______

______

4 Discuss the significance of decreasing artesian pressure and flow rates.

______

______

8.3 Aboriginal water use in the Barmah Forest

Knowledge and understanding • Case study

 **verbal–linguistic • visual–spatial**

The Barmah–Millewa Forest stretches across the border of New South Wales and Victoria. The Victorian section of the forest is the Barmah Forest. Study the information below and then answer the questions that follow.

The Barmah Forest

Natural flooding in the Barmah Forest occurred in the winter and spring, with dry periods during summer and autumn. However, since human intervention in the flow of the Murray River, water from winter and spring flows is stored in reservoirs. This means that the river is comparatively low at a time when it was usually and naturally high. The effect of this change in water patterns is a gradual deterioration in the condition of the forest ecosystem. This can be seen in:

- changes to the type of plants found in the rushlands, grasslands and forest
- reductions in fish and waterbird numbers and their breeding habits
- altered wetland hydrology
- reduction in species diversity
- poor tree health and quality
- decreased tree growth rate.

The Yorta Yorta approach to water management

As a matter of general policy, the Yorta Yorta people favour a water regime system which mirrors natural wetting and drying regimes, including reinstatement (restoring) of seasonal flood events of sufficient extent and duration. The Yorta Yorta people would like to maintain a minimum standard of water quality and to reduce the level of drainage of nutrients and herbicides into waters within the claimed area and the reinstatement of traditional cultural flows for cultural practices.

Water is not only essential for the continuation of Yorta Yorta culture and traditional rights, but is also important for the replenishment (restoring) of natural resources and the survival of the ancestral lands themselves. In this context the Yorta Yorta people don't make any distinction between water and land but see them as one whole system. Monica Morgan, who is a Yorta Yorta woman, expresses her concerns for a more equitable water management system by saying that 'there is a place for Indigenous People in the management of our waters just as there is for all Australians.'

Source: Relationship between Land Water & Yorta Yorta Occupation Paper prepared by Dr Wayne Atkinson for Yorta Yorta Native Title Claim, 1997

8.3.1 Barmah–Millewa Forest

1 Describe the pattern of water courses that run through the Barmah-Millewa Forest as shown in Figure 8.3.1.

2 Describe the natural pattern of flooding in the Barmah Forest.

3 Describe the impact of human intervention on the Barmah Forest.

4 Summarise the main approach to water management by the Yorta Yorta people.

5 Provide evidence which explains why the Yorta Yorta people see 'the water and land as one whole system'.

8.4 Irrigating Australia

Geographical skills

 logical–mathematical • visual–spatial

1 Complete Table 8.4.1 by calculating the percentage of agricultural land that is irrigated for all Australia and for each state/territory. This can be calculated from the area irrigated for each state divided by Australia's whole irrigated area, multiplied by 100.

	Area of agricultural land (ha)	Area irrigated (ha)	Volume applied (ML)	Application rate (ML/ha)	Percentage of Australia's irrigated land (%)
Australia	409 672 625	1 962 569	6 645 375	3.4	
NSW	58 326 346	674 064	2 745 896	4.1	
VIC.	12 625 915	487 368	1 134 701	2.3	
QLD	139 834 696	474 844	1 693 994	3.6	
SA	52 786 439	180 533	621 308	3.4	
WA	88 715 236	55 170	253 759	4.6	
TAS.	1 654 943	84 216	172 709	2.1	
NT	55 670 764	6 181	22 713	3.7	
ACT	^58 286	193	293	1.5	

Source: Australian Bureau of Statistics

8.4.1 Irrigation activity, by state and territory, 2010–11

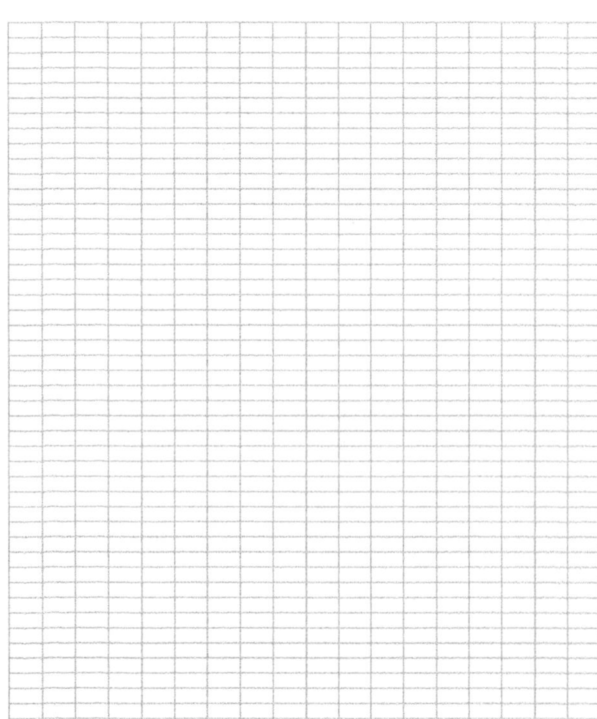

2 Construct a column graph of the volume of water used for irrigation by each Australian state/territory in 2010–2011.

3 Use Table 8.4.1 to help you complete the following sentence.

The application rate measures the amount of

__________________ used in irrigation for

every hectare of __________________ that

is to be irrigated.

4 Is the state with the highest volume of water used in irrigation also the one with the highest application rate? Use data from Table 8.4.1 to support your answer.

____

____

9.1 Water scarcity

Geographical skills

verbal–linguistic • visual–spatial

Study Figure 9.1.1 and answer the question that follows.

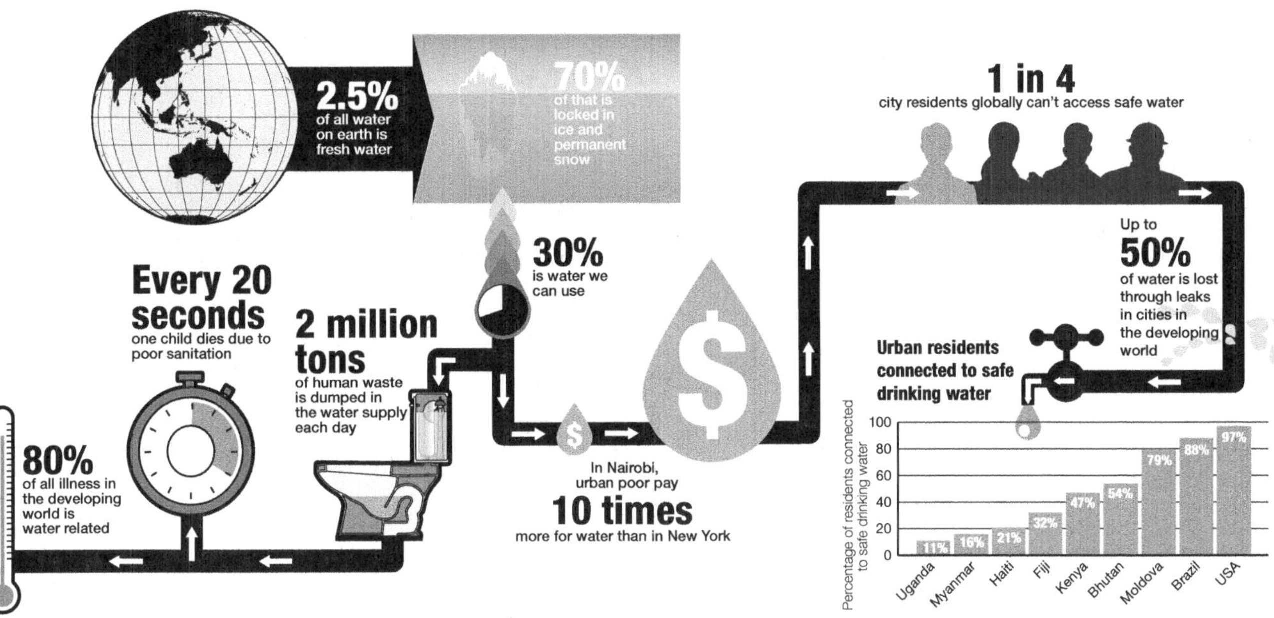

9.1.1 Facts about drinking water

1 Using Figure 9.1.1, write a story board for a television advertisement informing people about the worldwide issue of water scarcity. Plan out how each scene of the advertisement will look, and what facts and figures will be shown.

Study Figure 9.1.2 and answer the question that follows.

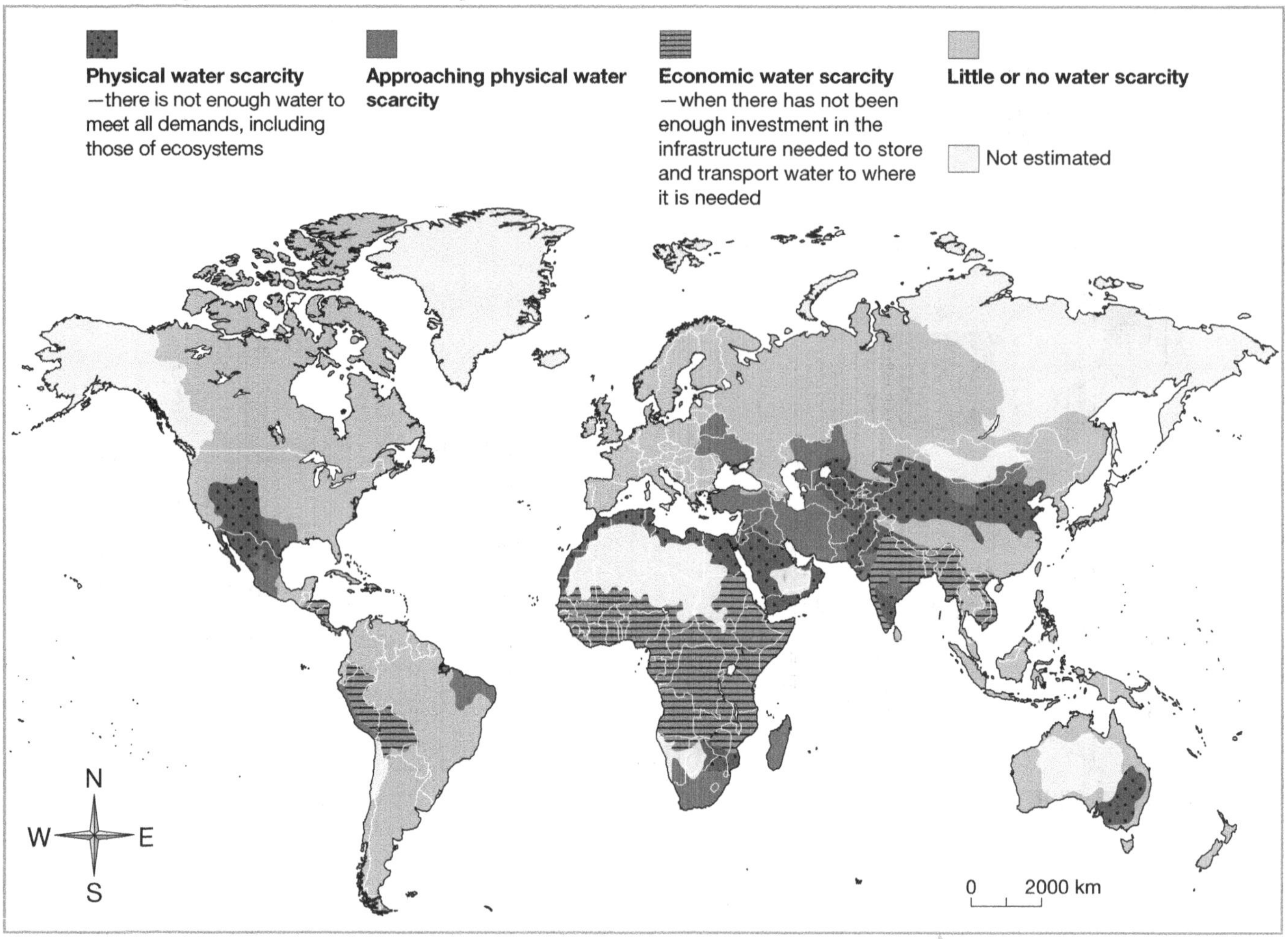

Physical water scarcity —there is not enough water to meet all demands, including those of ecosystems

Approaching physical water scarcity

Economic water scarcity —when there has not been enough investment in the infrastructure needed to store and transport water to where it is needed

Little or no water scarcity

Not estimated

N
W E
S

0 2000 km

9.1.2 Areas of water scarcity

2 Using Figure 9.1.2 and an atlas, discuss the spatial pattern of water scarcity in Africa compared with that in North America. In your response, describe the location of areas with water scarcity and areas with little or no water scarcity.

Pattern of water scarcity in Africa	
Pattern of water scarcity in North America	
Comparison	

Read the following article and answer the questions below.

South Africa is approaching water scarcity

The clock is ticking for South Africa's stretched water supply, and in another five years demand will have caught up with supply, according to a top official.

South Africa is constantly water-stressed. Although growth has slowed, an expanding economy, a growing population, and increased evaporation caused by climate change are combining to put additional pressures on water resources.

Yet leading experts at the conference said the situation could be addressed if the country reduced demand and improved water quality to enable reuse.

A paper by the World Wide Fund for Nature (WWF) said South Africa's water surplus had been dangerously low since at least 2000—four years after the country began buying bulk water from the multi-dam Lesotho Highlands Water Project, built on the Senqunyane River in neighbouring Lesotho.

Experts said the quality and quantity of the water supply should be better managed, and called for more investment.

Anthony Turton, who now works as a water management consultant, predicted that South Africa would soon have to start reusing effluent (sewage), which would entail rebuilding equipment, with waste treatment plants a priority.

Water treatment plants would have to produce effluent clean enough for reuse in the industrial sector, for example switching to buying cheaper, recycled water for cooling plants, he said.

…

Turton and others have also begun to conclude that if water could be stored in underground human-made aquifers, it could save a vast quantity of water from evaporation annually.

The government should crack down on hundreds of farmers who used water illegally from the Vaal River, 100 km south of Johannesburg, which supplies the city. The department of water affairs has established a unit, known as the 'Blue Scorpions', to police illegal bulk water use.

Source: IRIN, 22 May 2009

3 Explain what is meant by 'in another five years demand will have caught up with supply'.

____

____

4 What approaches have been suggested by experts to address water security issues in South Africa?

____

____

____

5 What methods have been suggested to reduce water loss from evaporation in South Africa?

____

____

____

____

9.2 Managing water resources

Knowledge and understanding • Geographical skills

verbal–linguistic • visual–spatial • logical–mathematical

Study Figure 9.2.1, which shows how humans interact with the water cycle, and answer the question that follows.

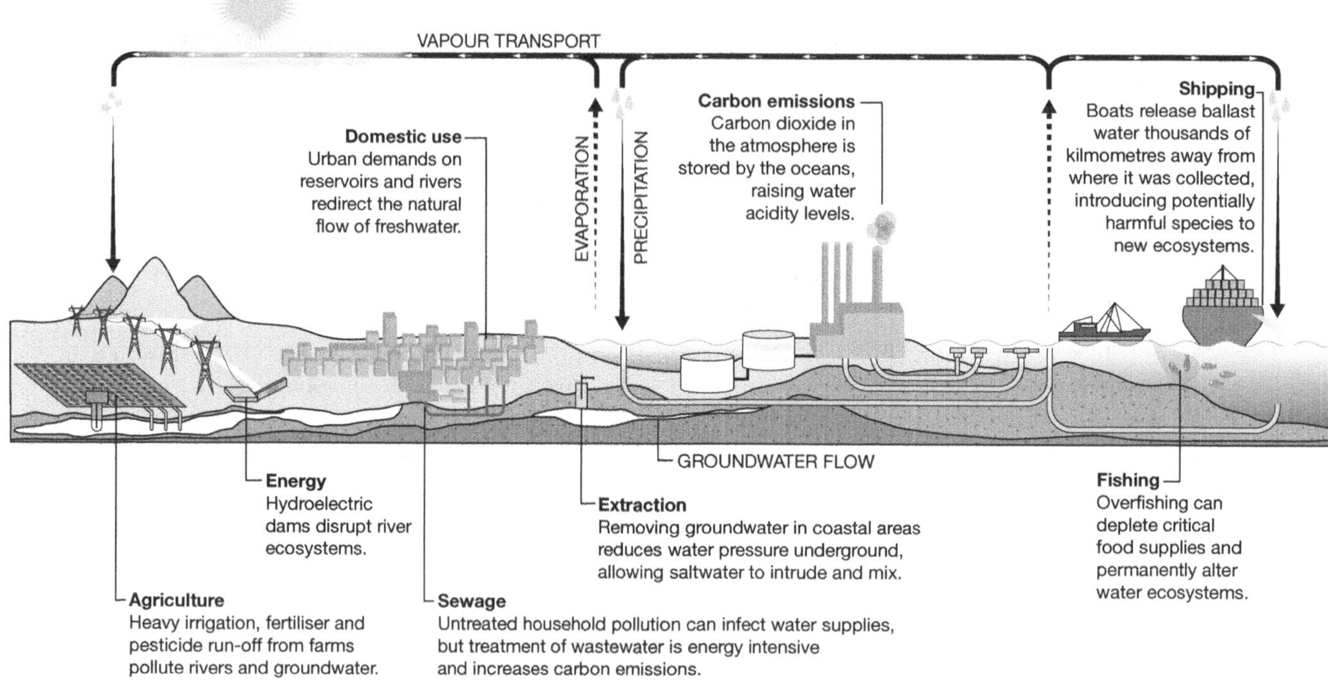

9.2.1 Human interaction with the water cycle

1 In your own words, provide four examples of how human activity impacts on water supply and water quality.

9.3 Dams and reservoirs

Geographical skills

 logical–mathematical • visual–spatial

Dam types

1 What is the main purpose of dams and reservoirs?

____

____

2 Work out the worldwide percentage of each dam type. Use the pie chart in Figure 9.3.1 and a protractor to calculate the answer.

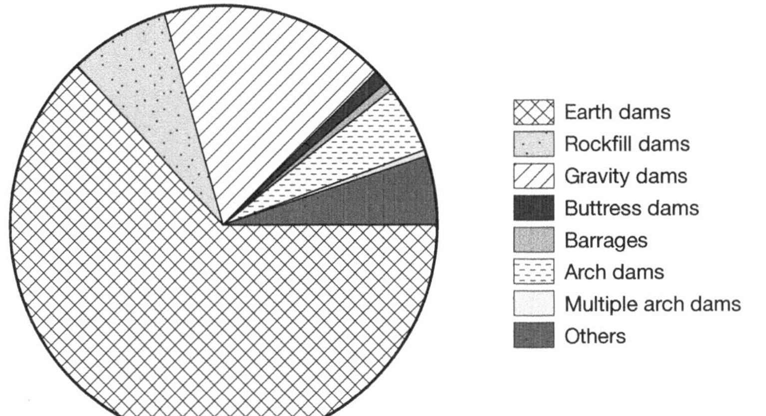

9.3.1

Percentage of dam types worldwide

a Earth dams ____________ **b** Rockfill dams ____________ **c** Gravity dams ____________

Purposes of dams

3 **a** Complete the table below by calculating the total number of single-purpose and multi-purpose dams.

b Calculate each percentage by dividing the number of dams for a purpose by the total number of dams and then multiplying by 100.

Purpose of dam	Number of single-purpose dams	Percentage
Irrigation	13468	
Hydropower	4914	
Water supply	3205	
Flood control	2603	
Recreation	1338	
Navigation/fish farming	151	
Others	1259	
TOTAL		

4 Complete the following statements using the data from your pie graph and the table in questions 2 and 3.

a Dams built for one purpose are most commonly built for ____________________.

b Eighteen per cent of all single-purpose dams are built for ____________________.

c __________ per cent of single-purpose dams are built for flood control.

5 Refer to Figure 9.3.2. Write a short speech arguing that the Gibe III Dam should not proceed and why.

____

____

____

____

____

____

____

____

____

____

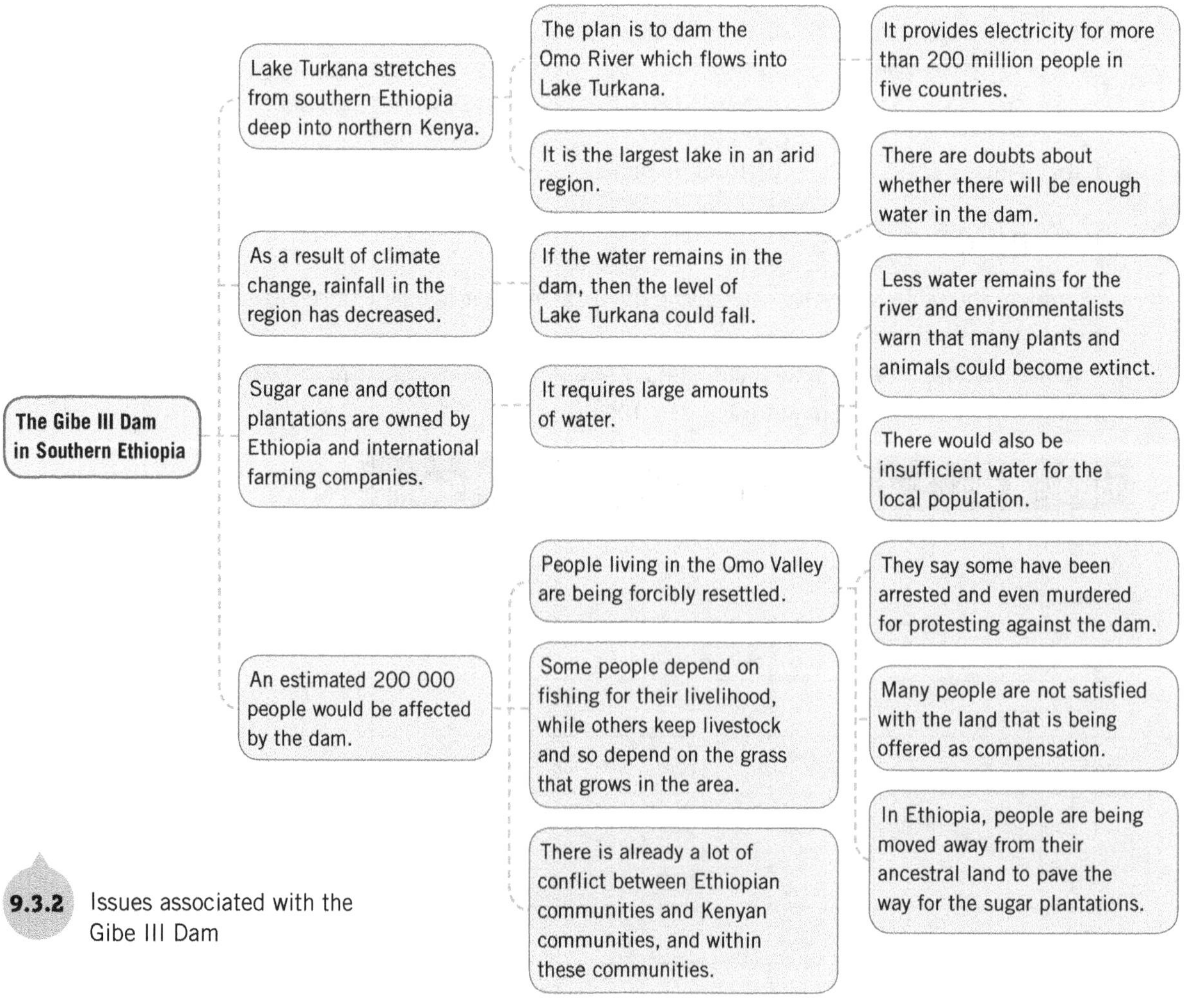

The Gibe III Dam in Southern Ethiopia

- Lake Turkana stretches from southern Ethiopia deep into northern Kenya.
  - The plan is to dam the Omo River which flows into Lake Turkana.
    - It provides electricity for more than 200 million people in five countries.
  - It is the largest lake in an arid region.
- As a result of climate change, rainfall in the region has decreased.
  - If the water remains in the dam, then the level of Lake Turkana could fall.
    - There are doubts about whether there will be enough water in the dam.
- Sugar cane and cotton plantations are owned by Ethiopia and international farming companies.
  - It requires large amounts of water.
    - Less water remains for the river and environmentalists warn that many plants and animals could become extinct.
    - There would also be insufficient water for the local population.
- An estimated 200 000 people would be affected by the dam.
  - People living in the Omo Valley are being forcibly resettled.
  - Some people depend on fishing for their livelihood, while others keep livestock and so depend on the grass that grows in the area.
  - There is already a lot of conflict between Ethiopian communities and Kenyan communities, and within these communities.
    - They say some have been arrested and even murdered for protesting against the dam.
    - Many people are not satisfied with the land that is being offered as compensation.
    - In Ethiopia, people are being moved away from their ancestral land to pave the way for the sugar plantations.

9.3.2 Issues associated with the Gibe III Dam

10.1 Weather hazards and disasters

Geographical skills

logical–mathematical • visual–spatial • verbal–linguistic

Refer to Table 10.1.1 to complete the following questions.

Weather hazard	Damage (US $ million)
Lightning	45.44
Tornado	9492.95
Thunderstorm	1327.08
Extreme cold	358.40
Extreme heat	9.99
Flooding	8216.90
Winter storm	169.35
Avalanche	0.06
Drought	2625.47
TOTAL	

10.1.1 United States damage from weather hazards for 2011

1 a Construct a column graph of damage (US $ million) for each type of weather hazard. Damage amounts go on the y (vertical) axis and type of weather hazard on the x (horizontal) axis. Make sure you include SALTS (scale, axes, legend, title, source).

Title: ____

b Complete Table 10.1.1 by calculating the total damage in US dollars.

2 Table 10.1.2 shows the number of deaths in the United States for each weather hazard in 2011.

a Complete Table 10.1.2 by calculating the total number of deaths.

b Calculate the percentage of deaths for each weather hazard and write them in the table. You can calculate the percentage by dividing the number of deaths for a weather hazard by the total number of deaths and multiplying by 100.

Weather hazard	Number of deaths	Percentage of deaths (%)	Sector of pie graph (°)
Lightning	26	2.7	9.7
Tornado	553		
Thunderstorm	56		20.9
Extreme cold	29	3.0	
Extreme heat	206	21.3	76.7
Flooding	69		
Winter storm	17		
Avalanche	11	1.1	0.4
Drought	0	0	0
Total number of deaths		100	N/A

10.1.2 United States weather fatalities in 2011

c From the percentages you calculated in question **b**, construct a pie chart of the number of deaths for each weather hazard. Use a protractor. Use the percentage of deaths column in Table 10.1.2 to calculate the sectors for your pie chart, with one per cent equaling 3.6°.

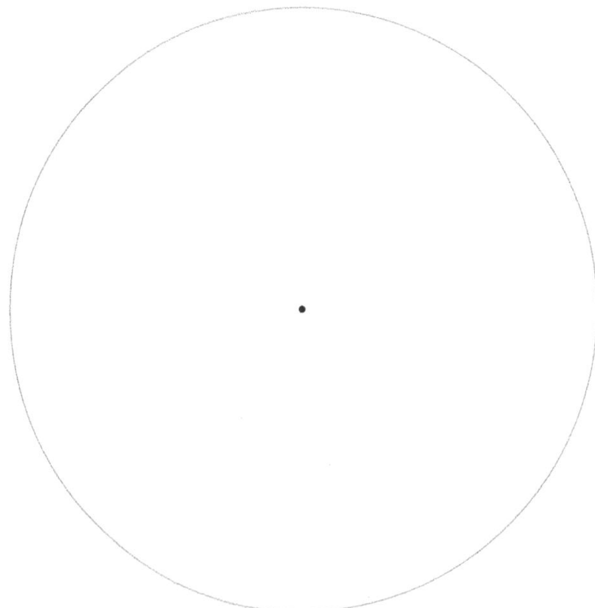

3 Use your answers from questions 1 and 2 to identify what you consider to be the most damaging weather hazard. Justify your answer.

10.2 Severe storms

Geographical understanding

verbal–linguistic • visual–spatial

Extension

1 Refer to Figure 10.2.1 to answer the following questions.

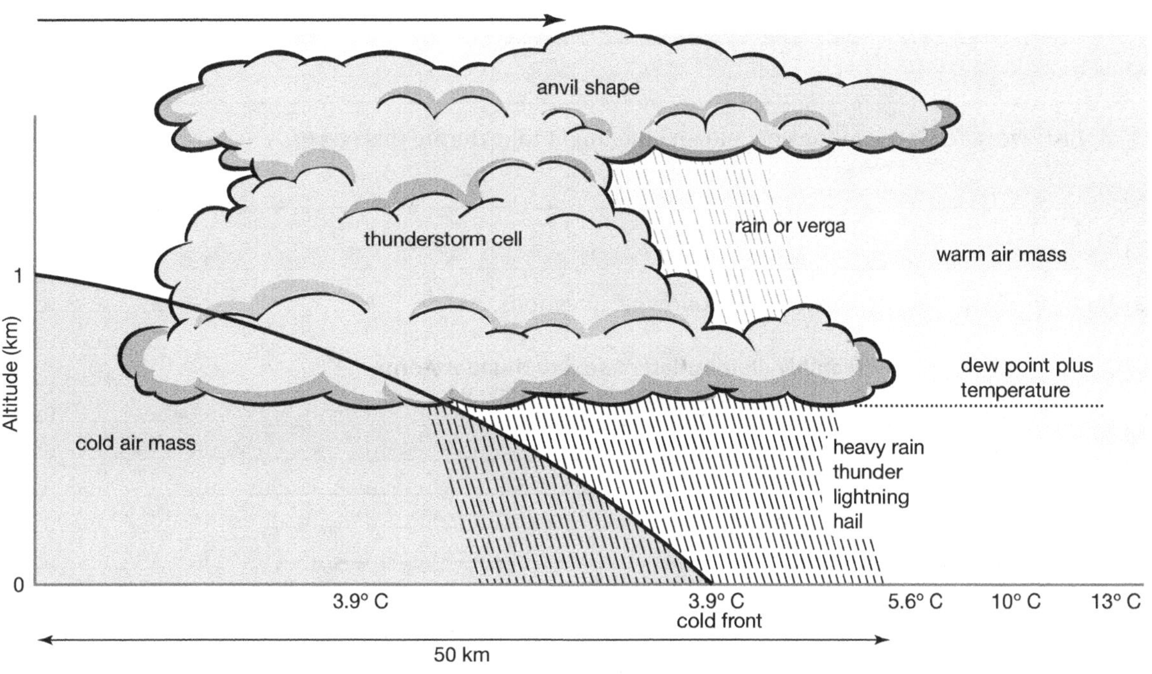

10.2.1 Formation of a thunderstorm when warm and cold air masses meet

a Describe the changes in temperature from just before a thunderstorm hits to just after it has passed overhead.

b Explain why a thunderstorm has formed in the situation shown in Figure 10.2.1.

c List four reasons why this storm may be considered a weather hazard.

2 a List two safety precautions people onboard boats might take during this storm.

b List two safety precautions people on land might take during this storm.

3 Refer to Figure 10.2.2 to answer the following questions.

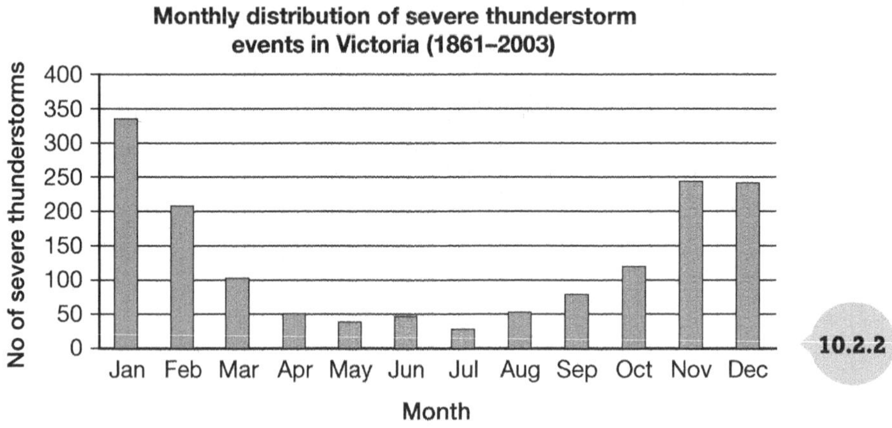

a Name the months that severe storms are most likely to occur in Victoria.

b What reason or reasons can you give for the yearly pattern of severe thunderstorms?

10.3 Tropical cyclones

Knowledge and understanding • Geographical skills

 **visual–spatial • logical–mathematical**

1 What is a tropical cyclone?

2 Describe the conditions needed for a tropical cyclone to develop.

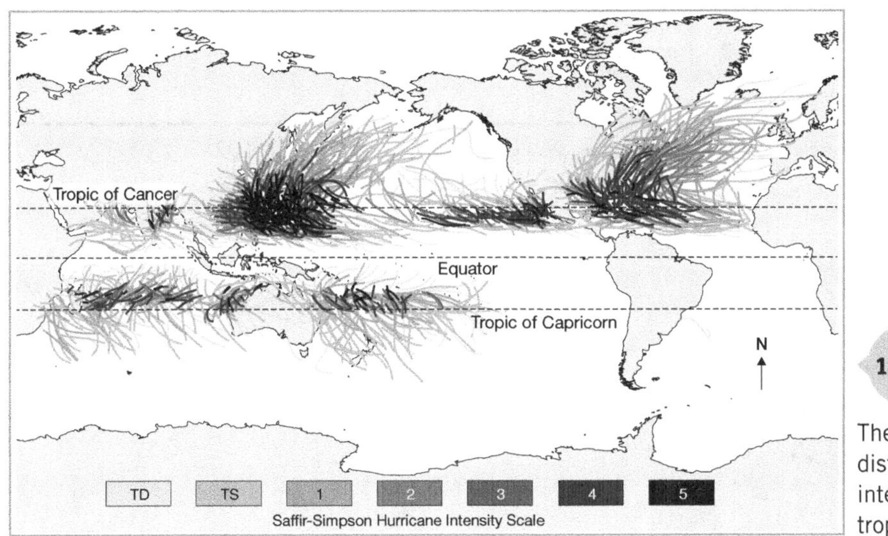

10.3.1 The worldwide distribution and intensities of tropical cyclones

3 a Study the Map in Figure 10.3.1. Complete the table below by writing in the frequency and intensity for each location.

Location	Tropical cyclone frequency: none/rare/common/very frequent	Tropical cyclone intensity using Saffir-Simpson Hurricane Intensity Scale: low(<1)/medium(2–3)/high(4–5)
South America		
Along Equator		
South Atlantic Ocean		
Central America and south-eastern United States (near Atlantic Ocean)		
West coast of Mexico (near Pacific Ocean)		
North-eastern Australia and the Pacific Islands		
East Asia (including Taiwan, Vietnam, Japan and the Philippines)		
North-western Western Australia		
North-eastern India and Bangladesh		
Southern Indian Ocean		

4 Table 10.3.2 shows a tropical cyclone category scale based on the Saffir-Simpson Intensity Scale. Use the information in this table to discuss the photographs below. Describe the level of damage, classify the cyclone as category 1–5 and give reasons why you classified the cyclone in that category.

Category	Wind speed	Wave height	Damage
1	Up to 125 km/h gales	1.2–1.6m	Slight damage. Damage to some crops, trees and caravans. Craft may drag moorings.
2	126–169 km/h destructive	1.7–2.5m	Significant damage. Minor house damage. Significant damage to signs, trees and caravans. Heavy damage to some crops. Risk of power failure. Small craft may break moorings.
3	170–224 km/h very destructive	2.6–3.7m	Structural damage. Some caravans destroyed. Power failures likely.
4	225–279 km/h very destructive	3.8–5.4m	Significant roofing loss and structural damage. Many caravans destroyed and blown away. Dangerous airborne debris. Widespread power failures.
5	Above 280km/h very destructive	More than 5.5 m	Almost total destruction and extreme danger. Houses flattened, cars overturned.

10.3.2 Tropical cyclone category scale, based on Saffir-Simpson Intensity Scale

Description of damage
Tropical cyclone category
Reasons

Description of damage
Tropical cyclone category
Reasons

Tropical Cyclone Yasi struck the northern Queensland coast very early on Thursday 3 February 2011. The Category 5 system caused an estimated 3.6 billion dollars damage to property and crops. The weather station on Willis Island, located 450 km east of the coastal city of Cairns, recorded the information displayed in Figure 10.3.3.

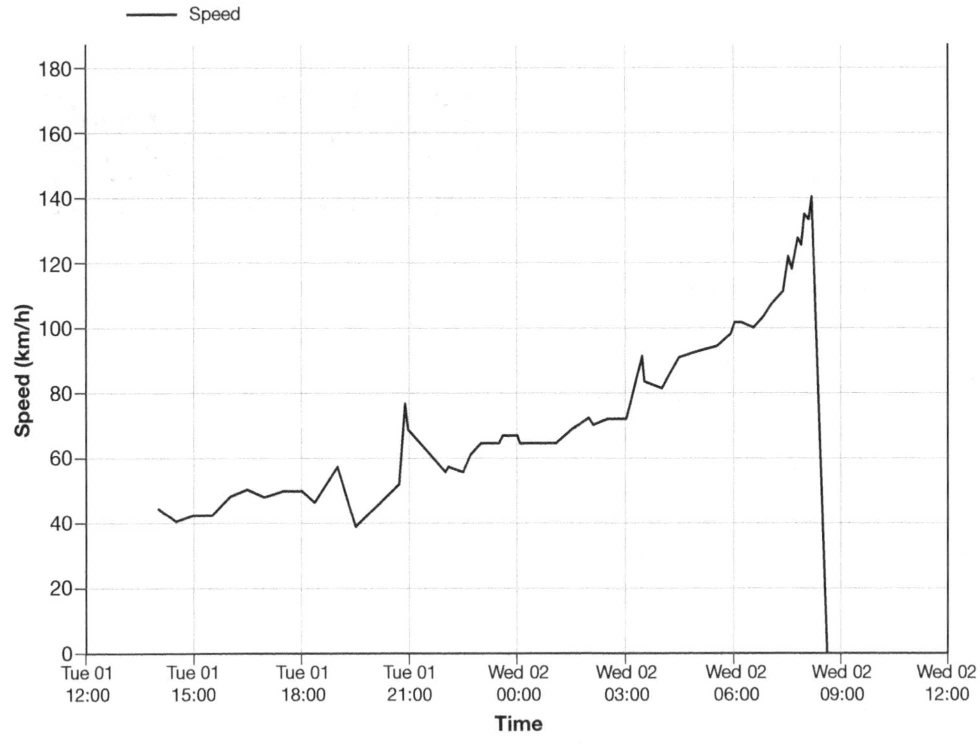

10.3.3

Wind speed over a 24-hour period from midday 1 February for Willis Island during Tropical Cyclone Yasi

5 Use the graph in Figure 10.3.3 to answer the following questions.

a Identify the speed of the wind at 00:00 on Wednesday 2 February 2011.

____

____

b Study the point that represents just after the cyclone reached its highest peak. Given that the weather station was not yet in the eye of the cyclone, what do you think might have happened to the instruments at the weather station to cause the drop seen on the graph?

____

____

____

i As the cyclone approached Willis Island, what was the highest recorded wind speed?

____

____

ii Did the tropical cyclone suddenly disappear? Explain your answer.

____

____

____

10.4 Tornadoes

Geographical skills

visual–spatial • logical–mathematical

1 a An average of 1253 tornadoes occur in the United States each year. The map of the United States in Figure 10.4.1 shows the average annual number of tornadoes for each state. Colour the map using the following instructions.

- Use six different tones of the same colour to represent the number ranges in the legend on the right. Use darker tones for higher number ranges and lighter tones for lower number ranges. For example, you could use white, light blue, baby blue, medium blue, navy blue and black.
- Once you have chosen your colours, colour the legend and corresponding states on the map.

Legend	
Number of tornadoes per year	**Colour**
0–10	
11–30	
31–50	
51–80	
81–100	
101+	

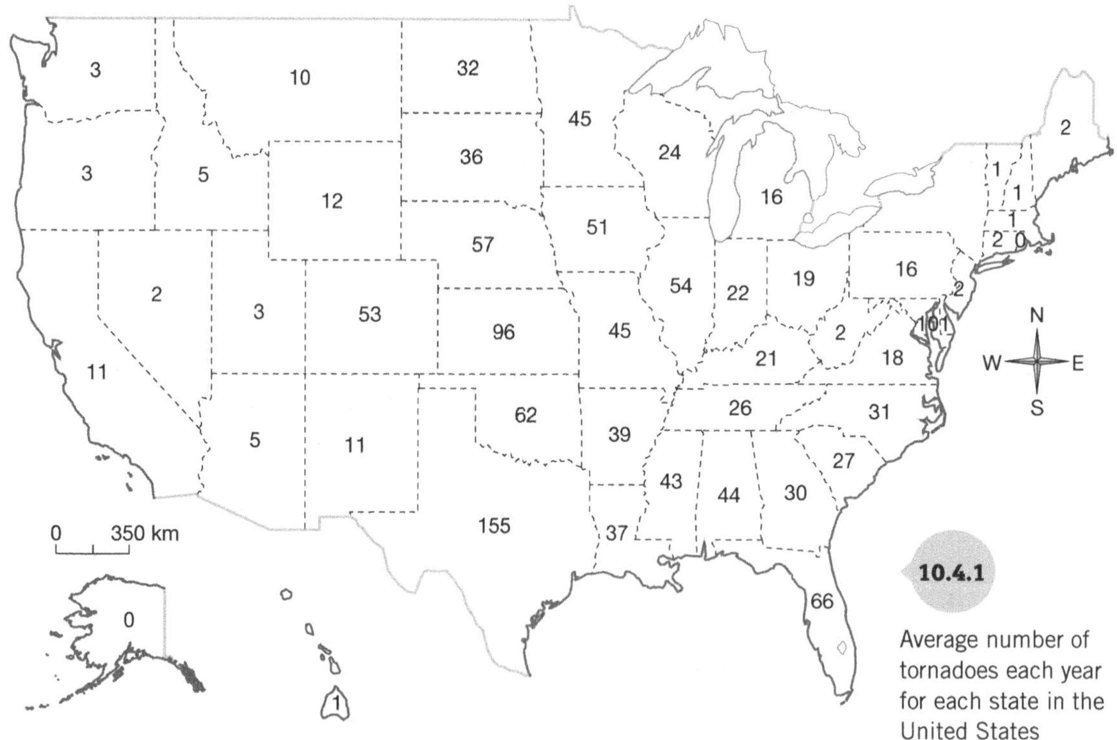

10.4.1 Average number of tornadoes each year for each state in the United States

b Define the term 'tornado alley'.

____

____

c Draw a line around the location of the tornado alley in Figure 10.4.1.

d Identify the type of map you have completed in Figure 10.4.1.

____

10.5 Flooding

Knowledge and understanding

 verbal–linguistic • visual–spatial

Extension

1 Complete the table below using the following descriptions of floods.

- lasting one or two days
- damaging; high risk to life and property
- sweeps away everything in its path: boulders, trees, buildings and bridges
- occurs where there are intense storms
- from one week up to months
- rivers in central and western New South Wales, Queensland and Western Australia
- little or no warning, reaching a peak in only a few minutes
- less than a day
- long period of build up
- stock loss, crop damage and extensive damage to road and rail links
- occurs quickly
- mountain headwaters of larger rivers and rivers, draining to the coast

Type of flood	Duration	Onset	Impact	Typical location
Slow-onset				
Rapid-onset				
Flash				

2 Study Figure 10.5.1 which shows the various ways human factors increase the risk of flooding.

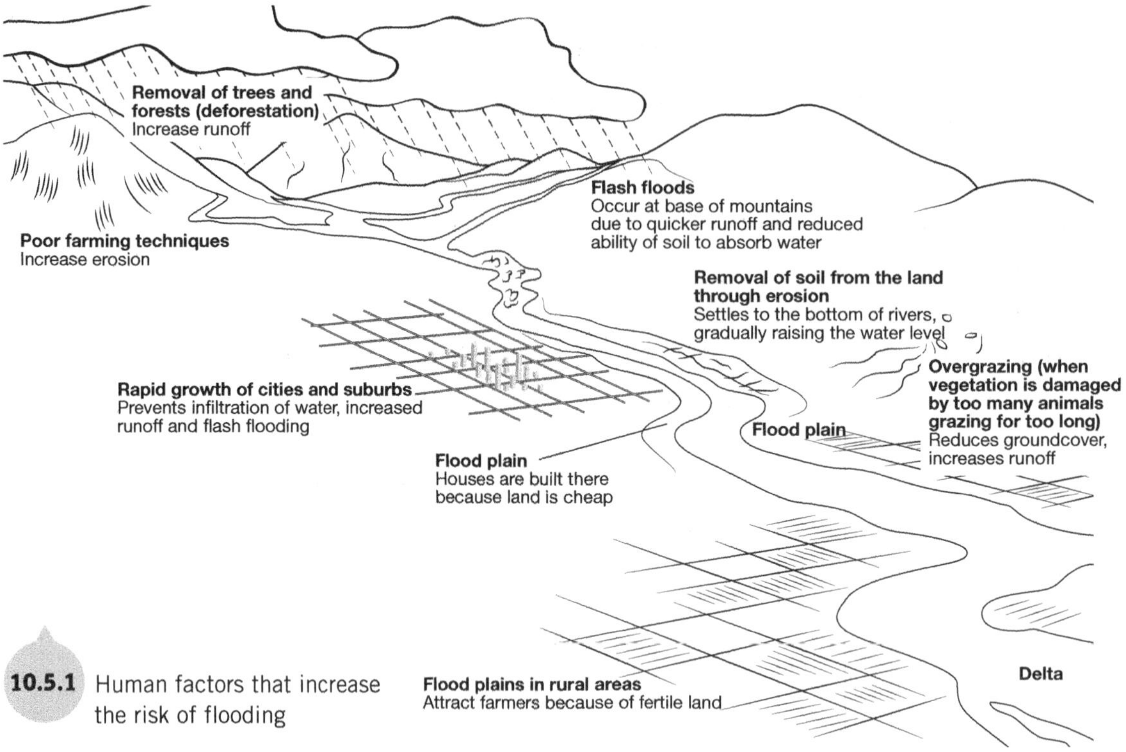

10.5.1 Human factors that increase the risk of flooding

a List three ways in which human factors can increase the risk of flooding.

____

____

____

b Propose some ways that floods could be managed in the area shown in Figure 10.5.1.

____

____

____

3 Use words from the box below to complete the paragraph about flash flooding.

deforestation	homeless	week	landslides	snow	tea	above
flooding	densely	110	collapsed	ground	sudden	heavy

Flash ____________ and landslides killed ____________ people and left 600 000 ____________. The ____________ populated and low-lying country has experienced almost a ____________ of torrential downpours. Most deaths were caused by ____________, although some were the result of ____________walls, lightning strikes and ____________ rushing floodwaters. In the rice and ____________ growing area of Sylhet, water rose a metre ____________ the houses, as residents scrambled into boats or higher ____________. Flooding in Bangladesh is usually caused by a combination of ____________ rainfall, melting ____________ and ____________ .

10.6 Droughts

Knowledge and understanding

 **verbal–linguistic**

1 Complete the table below by classifying the impacts of drought as economic (Ec), social (S) or environmental (En). Note that some impacts can be classified in more than one category.

Impact	Type of impact
Native grasses die off	
Topsoil loss through wind erosion	
Crop failures	
Anxiety and stress for farmers	
Soil becomes infertile	
Death of livestock	
Reduced value of livestock	
Family/relationship tensions	
Land degradation	
Loss of agricultural earnings and increased debt	
Increase in suicide rate	
Loss of biodiversity	
Increased inequality between rural and urban sectors	
Reduced economic growth	
Increased risk of extinction for endangered plants and animals	
Weed invasion	
Increased food prices	
Reduced food quality	
Build-up of combustible fuel	
Increased workload of farmers	
Increased frequency of bushfires	
Demise of smaller rural towns	
Banning of recreational activities on dams and reservoirs	

10.7 Definitions: Hazards

Knowledge and understanding

 verbal–linguistic • visual–spatial

1 The diagram below shows the six natural hazards. Label each description with the correct term from the box.

drought	tropical cyclone	tornado
heatwave	flash flooding	storm surge

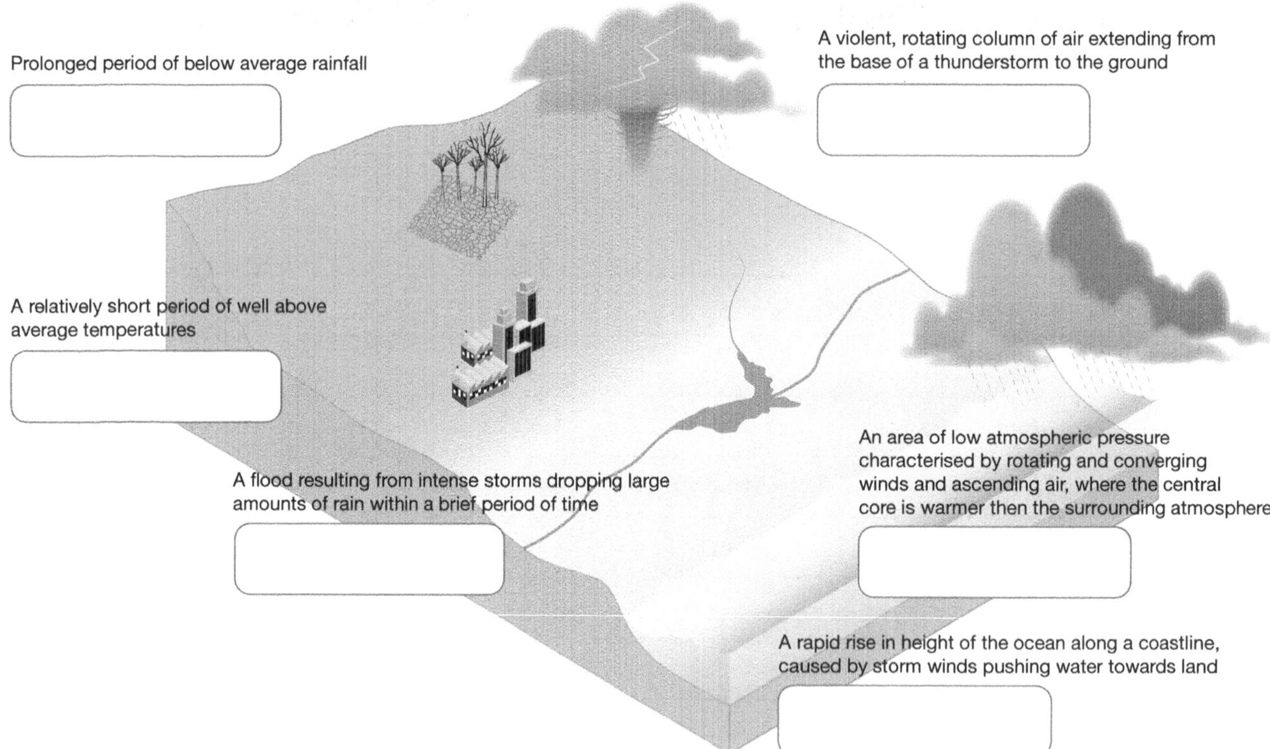

2 Match the following terms to their correct definitions.

Term	Definition
desertification	overheating of the body
flood mitigation	a flood that may last for one or more weeks
hyperthermia	the result when natural hazards impact on people
natural disaster	the expansion of deserts due to overgrazing, soil erosion, climate change or prolonged drought
natural hazard	actions taken to avoid, minimise or rectify the impacts of flooding
slow-onset flood	an event in the biophysical environment that is destructive to human life and property

11.1 Perception and use of places and spaces

Knowledge and understanding • Geographical skills

mi visual–spatial • interpersonal • intrapersonal

1 Your family is visiting Yellowstone National Park. You will also be taking a relative who has to use a wheelchair. Use the map in Figure 11.1.1 to plan where you will go with your relative. Include as much detail as possible.

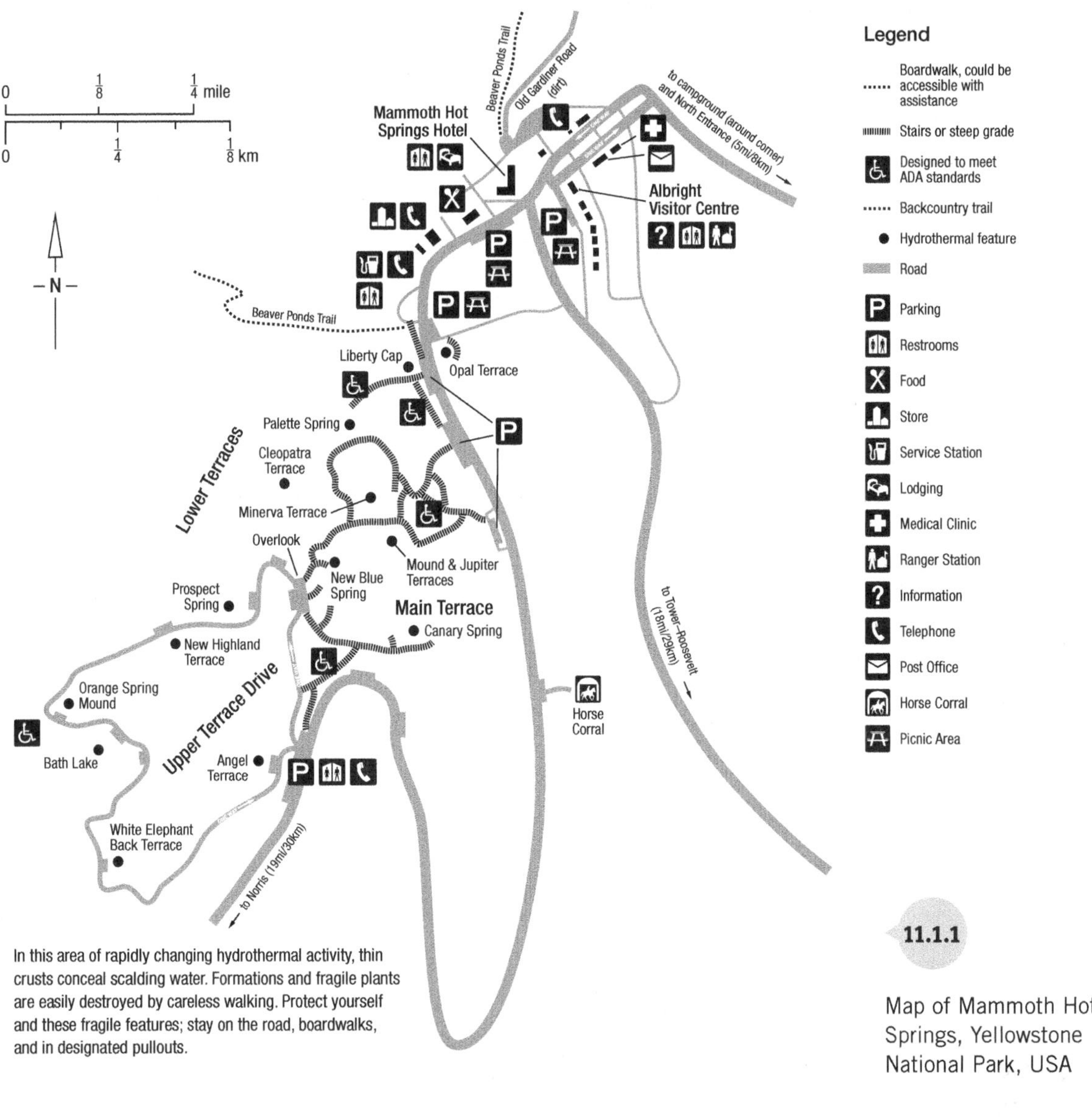

11.1.1

Map of Mammoth Hot Springs, Yellowstone National Park, USA

11.1.2

Ipanema Beach, Rio de Janeiro, Brazil

2 Describe your perceptions about Ipanema Beach based on the photograph.

____

____

____

3 What things can you not perceive about the location by just looking at the photograph?

____

____

4 Consider the following groups of people and outline how they might perceive the location in Figure 11.1.2.

a disabled person in a wheelchair: ____

b vision-impaired person: ____

c parent with young children: ____

d elderly person: ____

11.1.3

Tokyo, Japan

5 On the oblique aerial photograph of Tokyo, identify and label the paths, edges, districts and nodes of the city.

11.2 Place making

Knowledge and understanding • Geographical skills

visual–spatial • interpersonal • intrapersonal

Study the diagram in Figure 11.2.1 and answer question 1.

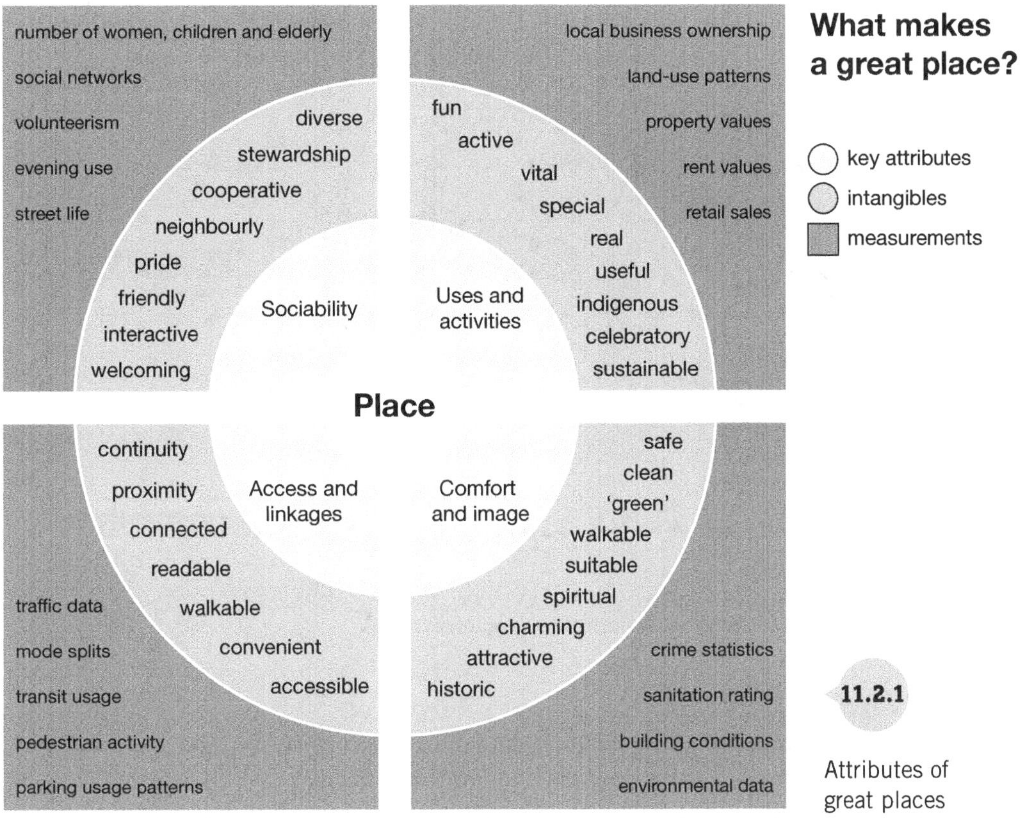

number of women, children and elderly
social networks
volunteerism
evening use
street life

diverse
stewardship
cooperative
neighbourly
pride
friendly
interactive
welcoming

Sociability

local business ownership
land-use patterns
property values
rent values
retail sales

fun
active
vital
special
real
useful
indigenous
celebratory
sustainable

Uses and activities

Place

continuity
proximity
connected
readable
walkable
convenient
accessible

Access and linkages

traffic data
mode splits
transit usage
pedestrian activity
parking usage patterns

safe
clean
'green'
walkable
suitable
spiritual
charming
attractive
historic

Comfort and image

crime statistics
sanitation rating
building conditions
environmental data

What makes a great place?

key attributes
intangibles
measurements

11.2.1 Attributes of great places

1 Think of a place that you would describe as great. Using the key attributes and intangibles, explain why.

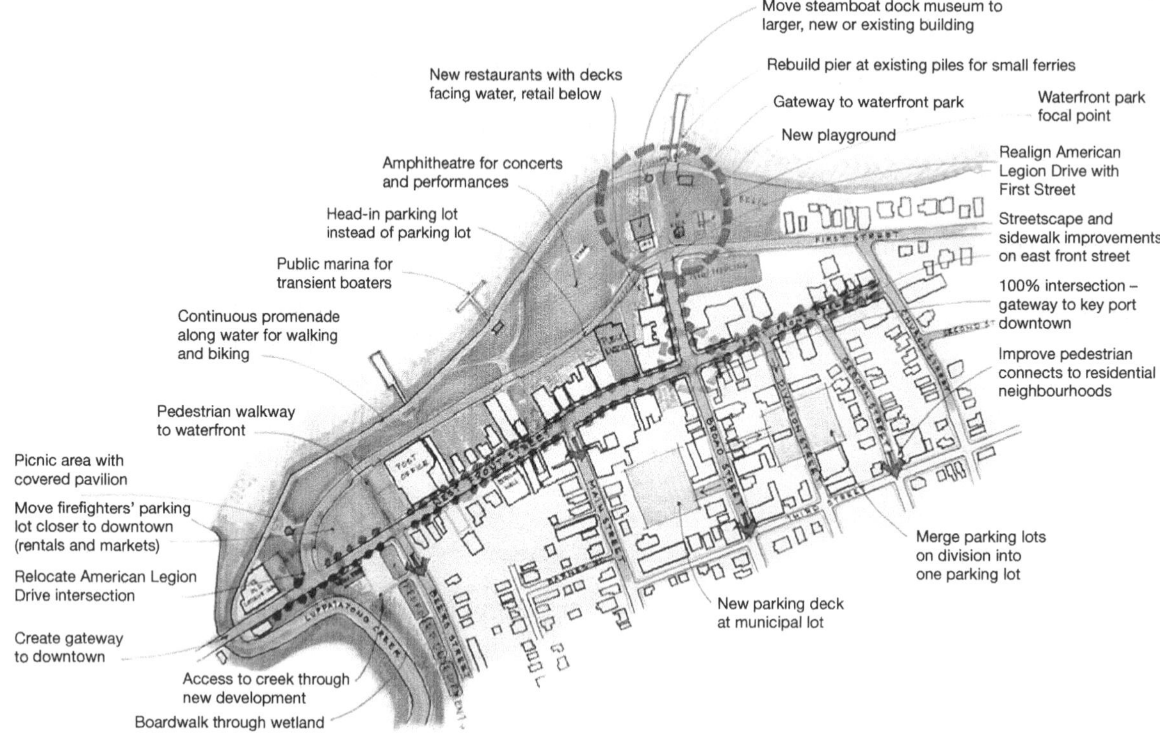

11.2.2 A place making plan devised for Keyport Waterfront, New Jersey, USA

Refer to the place making plan in Figure 11.2.2.

2 In the table below, rewrite four annotations that you think will significantly improve Keyport Waterfront. Explain why you think it will help create an improved place.

Annotation	Why it helps create an improved place

Territoriality of a place refers to the attachment of individuals or groups to a specific location. The location provides security and a physical expression of identity.

11.2.3 Street scene, India

3 The street in Figure 11.2.3 is being used by many groups of people. In the table below, identify each of the groups and state how you think they express ownership of the space.

Groups	How they express ownership of the space

4 Explain the potential territorial conflict between these groups.

11.2.4 The Townsville rodeo

The expression of territoriality can lead to distinctions between insiders and outsiders. The insiders of a place are those with common characteristics and shape the place to reflect what they perceive it should be like. Outsiders are people who visit on rare occasions or not at all.

5 What might distinguish the insiders in Figure 11.2.4 at this location?

6 How might outsiders appear to the insiders in Figure 11.2.4?

11.3 Definitions: Place making

Knowledge and understanding

 **verbal–linguistic**

Use the list of words in the box to complete the sentences below.

sense of place	surrounding	perception	meanings	space	construction
place	perception	packaging	places	linking	virtual	attachment
excluded	representations	reinterpreting	physical	environment	human rights

1 Mental maps are ____________________ of what people know about the locations and characteristics of places at a variety of scales.

2 ____________________ is the identification and interpretation of sensory information in order to represent and understand the ____________________.

3 Personal ____________________ is the area ____________________ a person which they regard as theirs.

4 ____________________ refers to specific areas of the earth's surface that have been given ____________________ which have been shaped by people.

5 Place making means the social and physical ____________________ of places.

6 Place marketing is the process of ____________________, designing, ____________________ and selling ____________________.

7 Place ____________________ is people's awareness of places and the particular opinions they have about them.

8 Place identity is the ____________________ of aspects of identity with place.

9 ____________________ refers to the characteristics that make a place special or unique, as well as to those that contribute a sense of human attachment and belonging.

10 Social justice is the protection of ____________________, ensuring fairness and rectifying social wrongs.

11 Social space is a ____________________ or ____________________ space such as a skate park or online social media, or other gathering place where people meet and interact with others.

12 Spatial inequality refers to places from which people feel ____________________ from some of the privileges available to others.

13 Territoriality is the ____________________ of individuals or groups to a specific location.

11.4 Global growth in tourism

Knowledge and understanding • Geographical skills

mi visual–spatial • verbal–linguistic • logical–mathematical

Growth of shark ecotourism

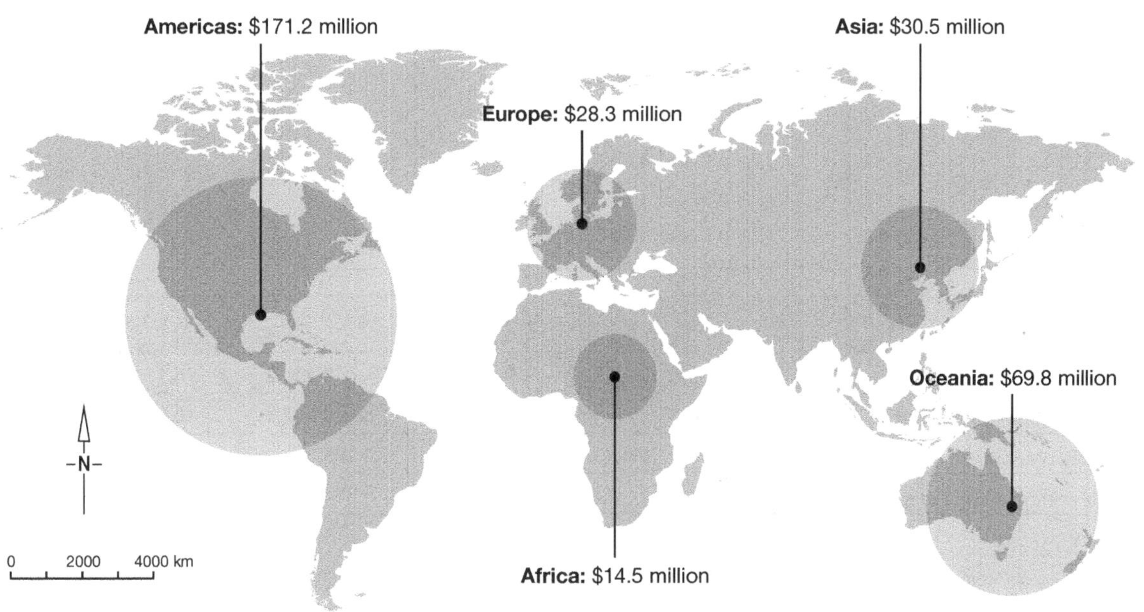

Source: University of British Colombia, 2013

11.4.1 Map of shark ecotourism

Sharks are keystone apex predators that have been keeping ocean food webs in balance for 450 million years. As a direct result of the largely unregulated shark fin fisheries and long-line bycatch, an estimated one-third of open-ocean shark species are now faced with extinction. Since all life is interconnected in the ocean, widespread shark depletion contributes to the loss of commercially important fish and shellfish species down the food chain, including key fisheries, such as tuna, that maintain the health of coral reefs and sea grass ecosystems.

1 Describe the present pattern of shark ecotourism.

____

____

2 Suggest how shark ecotourism could address the threat of widespread shark extinction.

____

____

____

3 List the potential advantages and disadvantages of shark ecotourism in the table below.

Shark ecotourism

Advantages	Disadvantages

World	2012 US$bn[1]	2012 % of total	2013 Growth[2]	US$bn[1]	2023 % of total	Growth[3]
Direct contribution to GDP	2,056.6	2.9	3.1	3,249.2	3.1	4.4
Total contribution to GDP	6,630.4	9.3	3.2	10,507.1	10.0	4.4
Direct contribution to employment[4]	101,118	3.4	1.2	125,288	3.7	2.0
Total contribution to employment[4]	261,394	8.7	'1.7	337,819	9.9	2.4
Visitor exports	1,243.0	5.4	3.1	1,934.8	4.8	4.2
Domestic spending	2,996.3	4.2	3.2	4,831.2	3.5	4.6
Leisure spending	3,222.1	2.2	3.2	5,196.0	2.3	4.6
Business spending	1,017.4	0.7	3.1	1,572.8	0.7	4.1
Capital investment	764.7	4.7	4.2	1,341.4	4.9	5.3

[1]2012 constant prices and exchange rates; [2]2013 real growth adjusted for inflation (%); [3]2013–23 annualised real growth adjusted for inflation (%); [4]'000 jobs

Source: World Travel & Tourism Council, 2013

11.4.2 Comparison of world tourism: 2012–2023

4 Consider each of the following claims a–h and decide if the data in Figure 11.4.2 supports (S) or does not support them (N).

a The amount of the total contribution of tourism to GDP will decrease from 2012 to 2023. ☐

b The growth in the total contribution of tourism to GDP will increase from 2013 to 2023. ☐

c Tourism is expected to result in an increase in jobs between 2012 and 2023. ☐

d Tourism is expected to provide a decrease in total contribution to employment between 2012 and 2023. ☐

e Visitor exports contributed the most to direct contribution to employment in 2012. ☐

f The amount of leisure spending in 2012 was greater than that of business spending in 2012. ☐

g Business spending will make a greater proportion of the total spending in tourism in 2023 compared with 2012. ☐

h Domestic spending on tourism will decrease because of global conflict between 2012 and 2013. ☐

11.5 Case study: The effects of tourism in Cambodia

Knowledge and understanding • Geographical skills

visual–spatial • verbal–linguistic • logical–mathematical

Hun Sen announces new airport

Prime Minister Hun Sen requisitioned 768 hectares of land on Saturday that will be used to build an international airport in Kampong Chhnang after 2025.

During presiding over a land title distribution ceremony in Kampong Chhnang province, the prime minister announced the expansion of the airport …

Ten days ago, Hun Sen said at the launch of the 2012–2020 Tourism Development Strategic Plan at Vimean Santepheap that the current airports cannot serve the increasing numbers of passengers expected to arrive in Cambodia …

Bun Rotha, Director of Phnom Penh International Airport, said by the time Kampong Chhnang Airport is completed, the site will be easily accessible from Phnom Penh because a new highway will be constructed specifically to serve the airport …

Source: Rann Reuy, *The Phnom Penh Post*, 29 October 2012, p. 7

Cambodian resorts on the increase

International tourist arrivals into Cambodia have soared in the past year, with arrivals into Siem Reap up 30 per cent in the first eight months of 2012 over the same period the previous year.

The tourist industry was hit hard by the global economic crisis in 2008 and 2009, but has recovered well in the last three years, with the Kingdom expecting over three million arrivals by the end of 2012.

With the ever-increasing numbers of tourist arrivals, experts say resorts are now becoming an exciting investment opportunity in Cambodia …

But with the increasing numbers of international visitors, there is more demand for western style boutique and resort style accommodation. According to global real estate giants CBRE, the first quarter of 2012 saw a 51 per cent increase in international visitors to Sihanoukville.

This increase has been partially driven by the introduction of the first flight into Sihanoukville from Siem Reap: part of the government's plan to connect the temples of the north with the beaches of the south. With 5741 passengers into Sihanoukville in the first six months of this year, it is proving to be popular, especially for more affluent travellers on short breaks, CBRE says …

Source: Rupert Winchester, *The Phnom Penh Post*, 4 October 2012

1 Read the two articles above then answer questions a and b. Write your answers in the diagrams.

a Predict possible positive and negative impacts of the new airport.

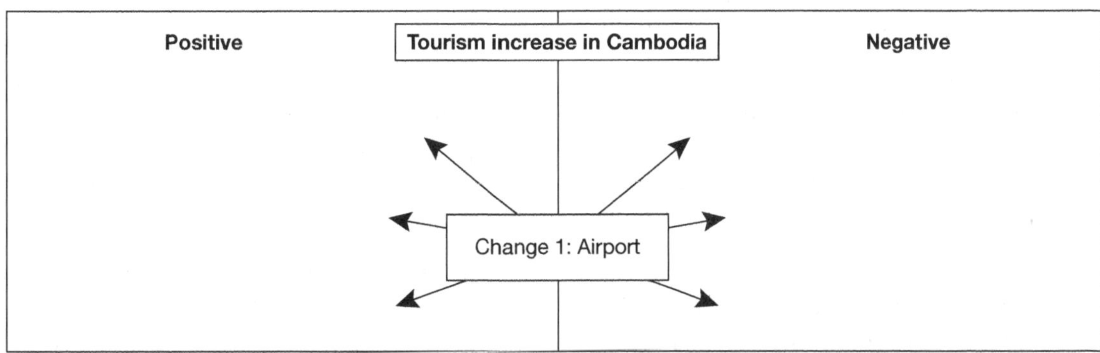

b Predict the possible positive and negative impacts of each of the changes.

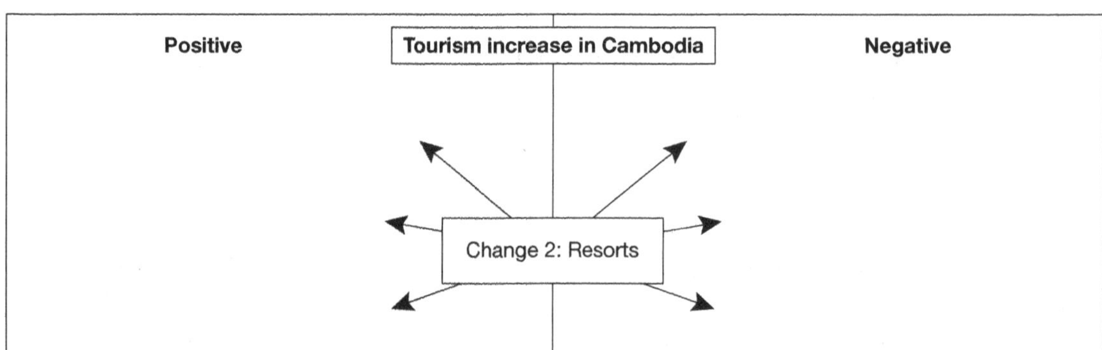

Little profit despite tourism growth

Despite a large increase in foreign tourists and tourism-generated revenue, Cambodia's poverty-stricken population is seeing very few of the profits, according to sources familiar with the industry.

The country's minister of tourism, Tong Khon, said recently that the number of foreign tourists visiting Cambodia in the first nine months of 2010 had risen nearly 15 per cent to 1.8 million from a year ago.

'The increase of tourists is due to our many important tourist attractions, such as the ancient ruins of the Angkor Wat complex, our beaches, and our eco-tourism, as well as attractions in the southwest regions of the country,' he said.

He also cited the rare freshwater dolphins of the Mekong River and mangrove forests of Koh Kong as large tourist draws, adding that improved road infrastructure, increased security, and economic recovery in Asia had all contributed to the growth of the Cambodian tourist industry.

A recent report from the Department of Tourism said income from tourism now accounts for 10 to 12 per cent of Cambodia's nearly US $11 billion Gross Domestic Product (GDP).

Countrymen benefit

But Van Peou, president of the Cambodia-based Informal Economics, said Cambodia's tourism industry is regulated in such a way that prevents the increased inflow of foreign tourist dollars from benefiting the population.

…

'The lives of local residents who deal with the tourists haven't improved much from the growth … The only income I can see [Cambodians earning] is from the sale of admittance tickets to see temples,' he said.

Son Chhay, a parliamentarian from the opposition Sam Rainsy Party, said that Asian tourists in particular, who make up the large majority of visitors to Cambodia, are more likely to spend their money on services linked to their home countries or that are provided by fellow nationals.

…

Chan Sophal, of the Association of Cambodian Economists, says very little of the nearly US $1 billion in profits from the tourist industry goes to the state.

'Angkor Wat has been run and controlled by a private company,' he said, referring to the Sokimex conglomerate, widely believed to be linked to members of the ruling Cambodian People's Party (CPP).

'All the airlines belong to foreign companies. Most tour companies, hotels, and restaurants are also owned by foreigners. Therefore the income does not go to Cambodia,' he said.

Little gain from increase

Workers in the tourism industry say that they have seen little gain from the increased number of visitors.

'This year it's harder for me to make money than the previous year because there are less European tourists and more competitors who quit working in the hotels to try this business,' said Ly Kitya, a motor tricycle driver in Siem Reap.

Another tricycle driver, who asked to remain anonymous, shared a similar observation.

'These days, most tourists join tours. They aren't the type of tourist traveling on their own that tend to want to see the Great Lake [Tonle Sap] or to go for a massage here and there.'

...

Chan Dy, manager of the Apsara Hotel in Siem Reap, agreed that visitors are far more likely to join tour groups, to the detriment of local guides and hawkers ... A source within the Cambodian Department of Tourism said that in the first nine months of the year, more than 400 000 travellers arrived via flights to Phnom Penh International Airport and nearly 500 000 to Siem Reap Airport—increases of 11 per cent and 17 per cent from 2009, respectively.

...

During a Sept. 27 speech commemorating World Tourism Day, Cambodian Prime Minister Hun Sen pledged to 'continue to accelerate [tourism] sector activity in order to help develop socioeconomically and work towards achieving the United Nations Millennium Development Goal.'

The goal to be achieved by 2015 includes, among other objectives, eradicating extreme poverty and developing a global partnership for development.

...

Source: Kim Peou for RFA's Khmer service, written in English by Joshua Lipes, 14 November 2010

2 Read the article then conduct a SWOT analysis of tourism in Cambodia—an investigation of the strengths (S), weaknesses (W), opportunities (O) and threats (T) that the issue presents.

S	W
O	**T**

11.6 Changing culture and places: Cycling to work

Knowledge and understanding • Geographical skills

mi visual–spatial • verbal–linguistic • logical–mathematical

11.6.1 A cycleway on one of the streets in Sydney's CBD

1 a Describe the facilities for cyclists evident in the photograph in Figure 11.6.1.

____

____

____

b Suggest reasons these cycling facilities have been provided by local councils and state governments.

____

____

____

____

	2001	2006	2011
Melbourne	12818	18912	25572
Adelaide	4605	6548	6451
Hobart	682	823	857
Perth	5650	6872	9240
ACT	3505	3758	4699
Brisbane	6895	8113	10373
Sydney	9279	10927	15608
Darwin	1567	1479	1678

Source: ABS Census

11.6.2 Bicycle commuters for Australian capital cities

2 Using the data in Figure 11.6.2, draw a line graph.

3 Compare the patterns of cycling to work in cities and suggest reasons for the differences.

4 To encourage more people cycling to work, list the facilities that councils and governments could provide.

-
-
-
-

11.7 Changing culture and places: Entertainment

Knowledge and understanding • Geographical skills

 **visual–spatial • verbal–linguistic**

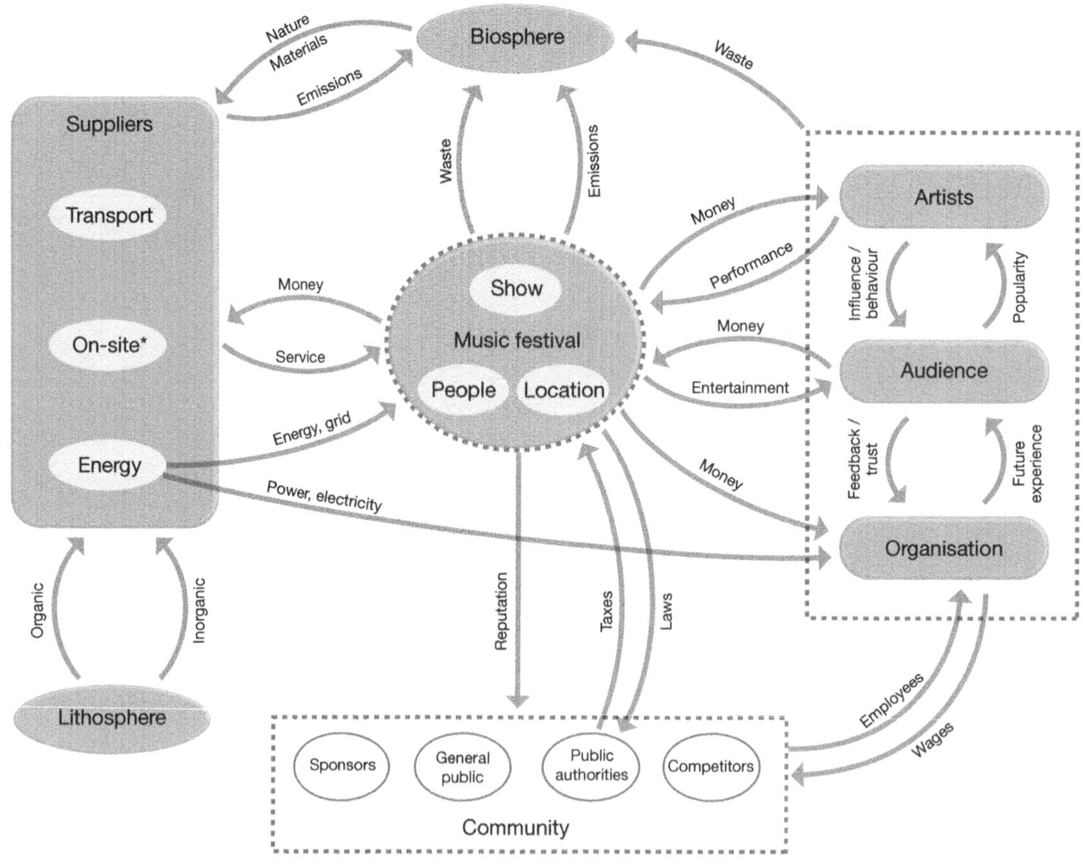

* The on-site suppliers include all the suppliers that are operating at the festival: vendors, security, infrastucture, cleaning, etc.

11.7.1 The links involved in a music festival

Study the diagram in Figure 11.7.1 then complete questions 1 and 2.

1 Simplify the diagram to focus on the inputs and outputs of a music festival. Complete the table below.

Inputs	Outputs
e.g. supply of transport service	e.g. payment for transport supply

2 Explain the relationship between a music festival and the place where they are held. (Consider the community, biosphere and local businesses of the place.)

____

____

____

____

____

The City of Plano, Texas has signed a contract for a two-day music concert to be held at the Oak Point Park and Nature Reserve. The City wants the festival to be similar to the Austin City Limits Music Festival, but smaller. There are different opinions about the festival. These include:

- The council wants to bring people into the park, which many consider a hidden gem.
- 20 000 people are expected to attend the music festival, bringing revenue to the city.
- The nature reserve includes native prairie land—much native prairie land in Texas has been destroyed.
- One of the concert sites will sit on top of rare native prairie land.
- Conservationists fear that this rare native prairie land will be destroyed by the festival.
- Conservationists want to use the native prairie land for educational purposes.

11.7.2 Austin City Limits Music Festival

Is there a music festival bubble?

If you haven't already, grab your feather earrings and fanny pack, because we're deep into summer music festival season … With April's mega-popular Coachella marking the unofficial start of the season, festivals across the country are now anticipating their own waves of crowds to turn up, scope out the best port-a-potty, and spend lots of money.

Festivals have long been popular—the 1724 Three Choirs Festival in Gloucester, I hear, was a rager—but their recent surge in popularity in America has much to do with their unique role as a cultural commodity. In a 2010 study from the University of Queensland, researchers polled young festival-goers, ages 18 to 29, about their experiences. Eighty-three per cent felt they came away more hopeful about the way things are in the world, while 91 per cent felt more positive about their lives. Festivals make attendees feel feelings, and they'll pay—and even wear an electronic wristband—to keep feeling them.

With festivals able to make local economies swell, more towns are trying to get a piece of the pie.

Today, concert tickets cover about 40 per cent of the 50 per cent drop in U.S. recorded music sales since 1999, as reported by *LA Weekly*. The country's most established festivals—Coachella, Lollapalooza, and Bonnaroo – each sell out with well over $20 million in ticket revenue. With profits high, festivals sometimes ensure their monopoly over a touring band by insisting on exclusivity clauses, often restricting the artist from playing at other public or private concerts within 300 miles [480 kilometres] of the festival for 180 days prior and 90 days past the event.

While it may be easy for festival-lovers to gripe at the apparent greediness of promoters—Coachella's single-day passes went for $90 in 2008; now a $349 full-festival pass is the cheapest option—festival revenue can also have a huge impact on hosting cities. August's Outside Lands brought 683 short-term, full-time jobs to San Francisco and infused $67 million into the local economy in 2011. SXSW, which also includes films and a speaker series, brought $167 million to Austin. And the pumping, glow-stick-loving Electric Daisy Carnival brought a smooth $207 million to Ibiza-wannabe Las Vegas.

With festivals able to make local economies swell, more towns are trying to get a piece of the pie. Napa's BottleRock just enjoyed a successful premier, and Monterey will be trying their luck this August with First City—although the California seaside town can already boast about its festival pedigree as home to the infamous Monterey Pop Festival, the first widely promoted and attended rock jubilee. But will the increase in festivals cause the bubble to burst? Despite the participants' festival adoration, the Queensland study also found that over-attendance detracts from emotional impact. It was only those who attended festivals once every few years who reported the highest levels of well-being outcomes.

Just don't forget your water bottle.

Source: Sarah Sloat, *Pacific Standard*, 13 June 2013

Read the two articles about music festivals and answer questions 3 and 4.

3 Create a mind map about the social, economic and environmental impacts of music festivals.

Music festivals

4 Prepare a speech you would give at a public meeting set up to debate the plans for a music festival at the city of Plano. Your 3-minute speech should clearly argue your position about whether or not the music festival should go ahead.

11.8 Geographical terminology: Tourism

Knowledge and understanding

 **verbal–linguistic**

Amy is writing a travel diary during her holiday overseas. Unfortunately, she has not used geographical terminology. Use the words in the list below to improve her use of geographical terminology. Write the appropriate word in the space above the appropriate phrase.

cultural diffusion	cultural group	culture	popular music	tourism
leisure	senior citizens	ecotourism	employment	commuters

Day 4

Even though I've been here for four days, I am still finding it hard to understand how things work in Tokyo. The Japanese have a very different complex mixture of origin, heritage, language, religion, customs and way of life to my own. Even the music that has wide appeal is new to me. I recognise some things like certain fast food chains and dress, particularly for the men. Clearly there has been a spread of cultural characteristics from other places to here.

Day 5

Today we took a tour up to Mt Fuji. We couldn't get to the top by coach and the tour guide said we would not have time. Still, the view we had from where we stopped was great. We overlooked a valley with a lake. Unfortunately, there were so many people that it was hard to stay in the best viewing spots for long. There were a lot of coaches with old people. Many Japanese were also making the most of the weekend. I believe that many Japanese work long hours and don't get much opportunity to do things during their free time.

Day 6

We took the train network to a few places. Although the signs are not in English, it is fairly clear how to get from place to place. Our biggest mistake was to try to leave really early in the morning. This was Dad's idea to make the most of the day. We were trying to leave with the rush of people travelling to work. Did you know that there are 63.59 million people in a job in Japan? No wonder the public transport is so crowded.

Days 7 and 8

Today we visited a little village near Tokyo that aims to encourage travel to wild places in ways that conserve the environment, now and in the future. We spent the night in a place called Gankoyama Tree House Village and learnt about Japanese forestry survival skills. (Maybe more about this in my blog later).

Day 9

This was our last full day in Japan. We spent most of the day wandering through Tokyo trying the food and shopping. Many places thrive on the activity of travel for recreational, leisure or business purposes. My impression is that Japanese society is quite similar throughout. There appears to be just one group of people with a similar heritage and language and ancestry. This seems a big contrast to my own city.

12.1 Transport connections

Knowledge and understanding • Geographical skills

 visual–spatial • verbal–linguistic

12.1.1

The Donghai Bridge has a total length of 32.5 kilometres and connects mainland Shanghai and the offshore Yangshan deep-water port in China.

1 Explain how transport technology may have helped improve connections between Yangshan and Shanghai.

2 Shanghai is a major Chinese city. Explain how the Donghai Bridge could be an advantage to both Shanghai and Yangshan.

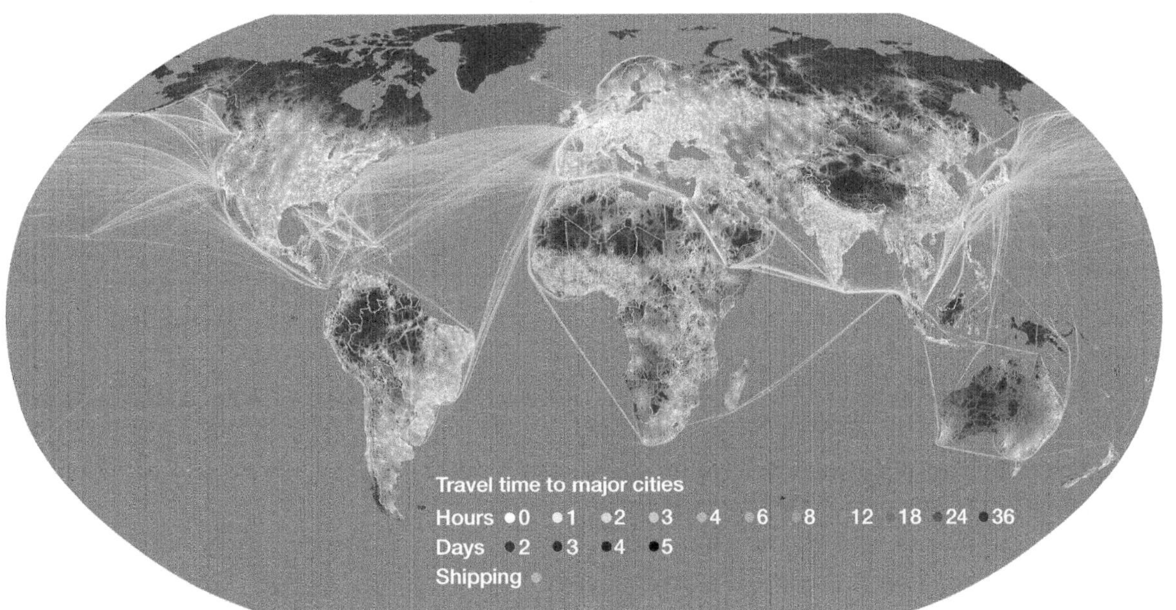

Source: European Commission's Joint Research Centre and the World Bank

12.1.2 World map showing remoteness index. The remoteness index is a measure of the time it takes to travel to the nearest major city.

3 Using the PQE method, describe the locations that are the most remote in the world according to the map.

- Pattern: ______________________________
____
- Quantity: ______________________________
____
- Exceptions: ______________________________
____

4 State regions of the world that are most connected via shipping.

____

5 Explain the advantages of shorter travel times to major cities.

____
____

Map showing major rail links throughout Australia

6 Explain why Mount Isa, Charleville (Qld) and Kalgoorlie (WA) might be considered less remote than Tambo (Qld) and Geraldton (WA) respectively.

____
____
____

12.2 ICT connections

Knowledge and understanding • Geographical skills

 **visual–spatial • verbal–linguistic • logical–mathematical**

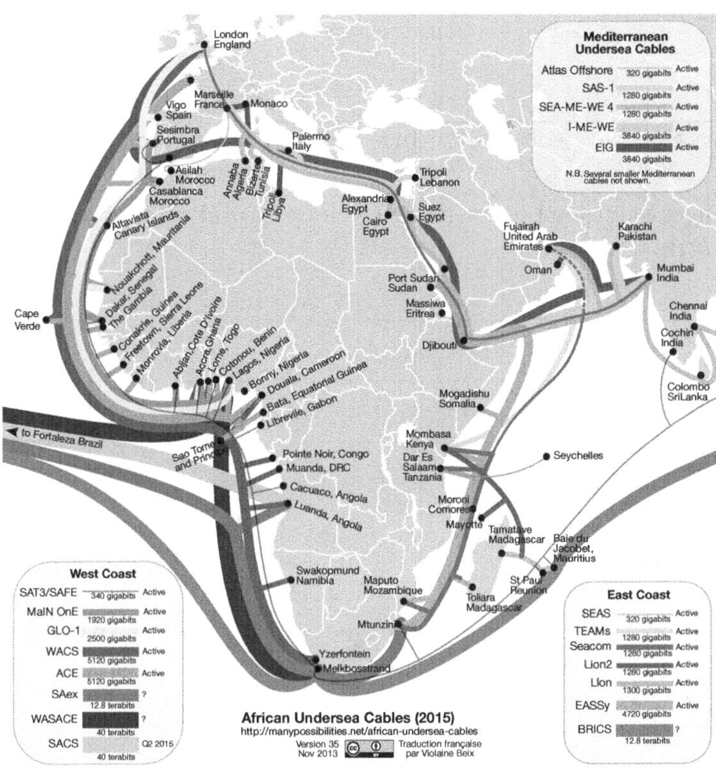

12.2.1 The impact of 2010 FIFA World Cup on telecommunications in Sub-Saharan Africa

Football and broadband

The telecommunications needs of the 2010 FIFA World Cup in South Africa helped bring a plethora of new international subsea cables to Sub-Saharan Africa.

1 What did the international subsea cables enable many Sub-Saharan countries to upgrade?

____

____

____

2 a Identify the countries that benefited from faster internet connections in 2012.

____

____

b Describe the spatial distribution of these countries.

____

____

c List the possible impacts this spatial distribution could have.

____

____

3 Explain how ICT connections could benefit these African countries in terms of:

a social media: ______________________________

____

b new industry and employment: ______________________________

____

c new ways of buying: ______________________________

____

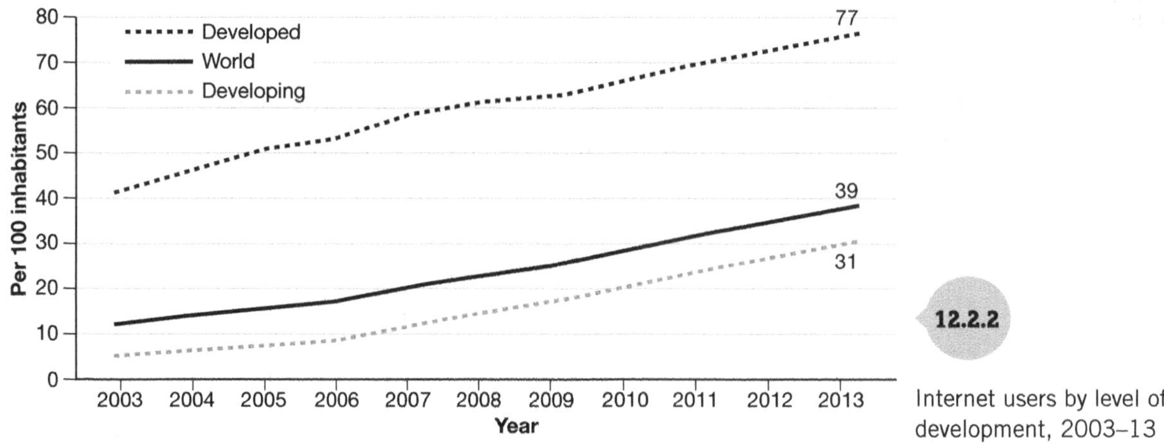

12.2.2 Internet users by level of development, 2003–13

4 Study Figure 12.2.2. Compare the changes in internet users in the developed world with those of the developing world, between 2003 and 2013.

____

____

____

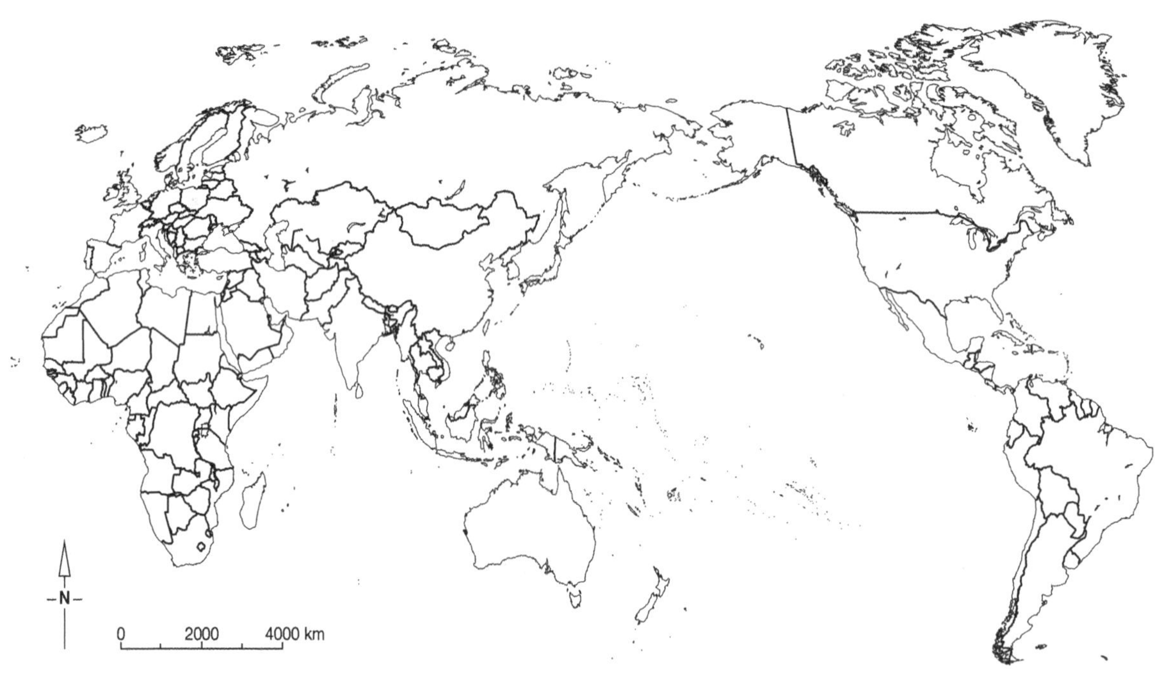

12.2.3 World map

5 Use the map of the world in Figure 12.2.3 to complete the following activities.

a Locate and label the places that you are regularly connected to via ICT.

b Connect a line from your home to these places, giving the line a colour according to the category of connection; for example, family, friends, entertainment, commercial, education etc.

c Write a statement about how important the ICT connections are to your life.

____

____

12.3 Trade connections

Knowledge and understanding • Geographical skills

verbal–linguistic • visual–spatial

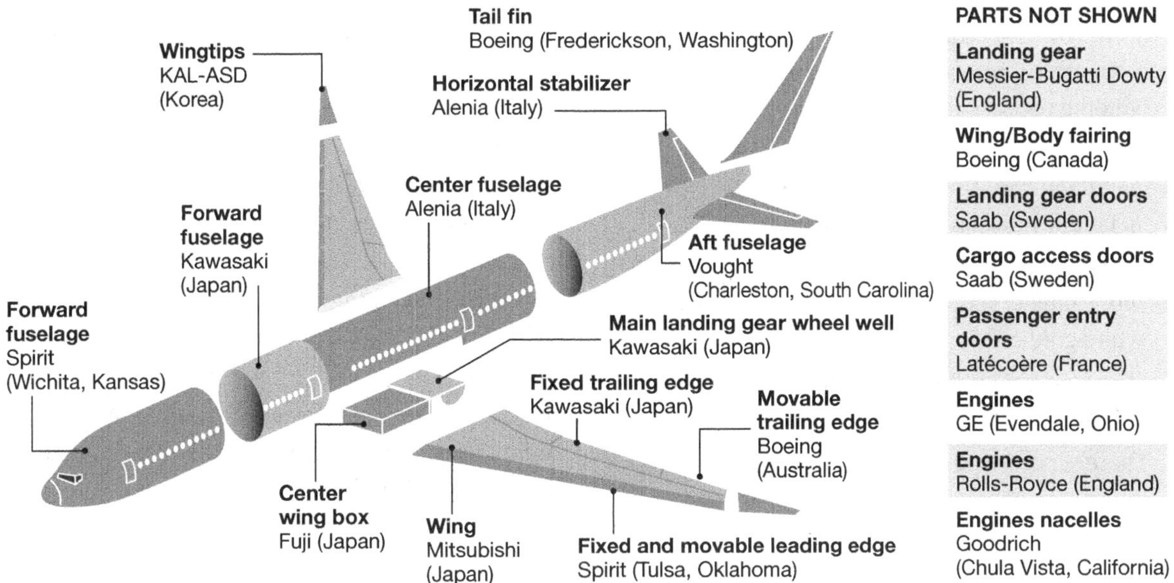

12.3.1 Trade links in the assembly of a Boeing 787

1 Using the trade links in Figure 12.3.1, mark all the countries and places associated with building a Boeing 787.

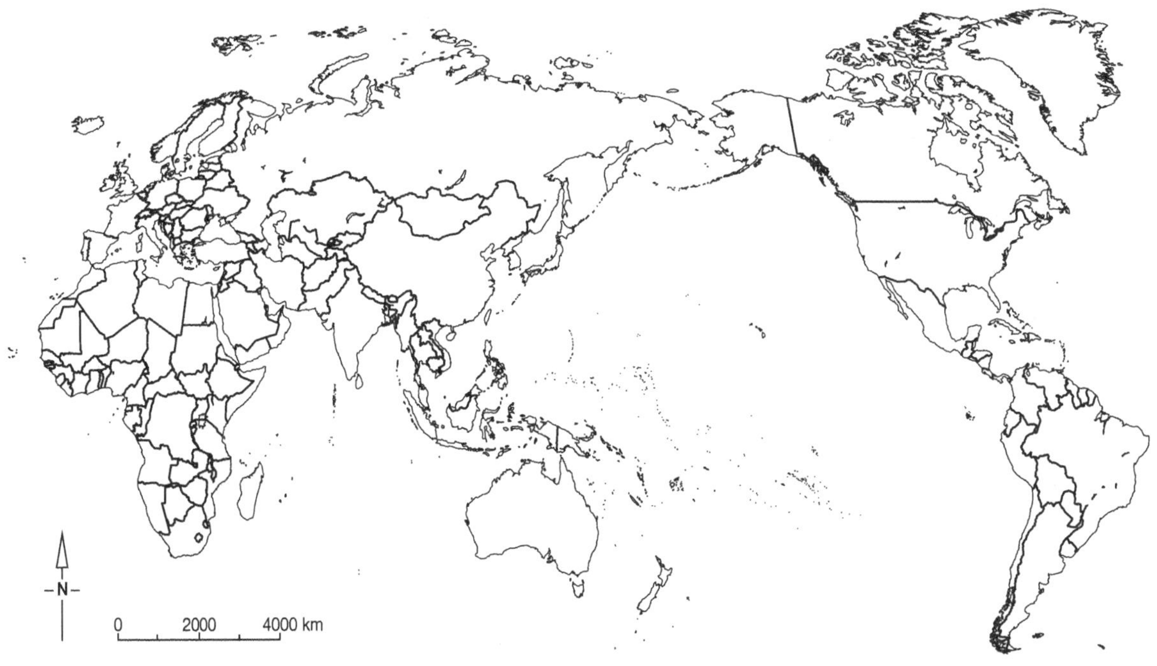

12.3.2 World map

White gold rush—Milk formula exports to China

Infant formula is in such hot demand in China that smugglers have been arrested trying to import large and small quantities of what's being termed 'white gold'.

Baby formula, or infant milk powder, is the only growth sector for food manufacturing in Australia, with start-up companies already building processing plants.

But food safety is top of mind for most manufacturers—particularly after China's melamine scandal of recent years.

As well, the lack of a Free Trade Agreement with China is seriously holding back Australia's 'white gold rush'.

Every year, more than 17 million babies are born in China.

There's a growing middle class, and as mothers return to work they want the best for their one child that money can buy.

Once upon a time, the Chinese upper classes used to employ wet nurses. These days the milk comes from a tin, called Australian Yummy, Gold Roo, or Joey's infant formula.

It doesn't come cheap—costing up to $70 a tin, $50 more per tin than in Australia.

Locally, as tomato processors close and canneries around the country struggle, infant milk formula is the only area of food manufacturing that is not on life support.

In the central Victorian town of Tatura, 400 people have good employment prospects, turning liquid milk into infant powder.

In vast stainless steel silos, vitamins, minerals, blended vegetable oils and extra protein are added to liquid milk which eventually becomes bulk infant formula.

...

In 2008, Chinese agents were caught adding the industrial chemical melamine to try to boost protein and profits; it killed six babies in China and hospitalised 300 000 with painful kidney stones ...

'In China 17 million babies are born every year, and in the first quarter of 2012 China imported more dairy products than for the entire year of 2007.

'So in five years, they've had enormous amount of growth in demand, particularly for imported dairy products.

Last year, Australia earned $20 million from infant milk powder sales to China. It's expected to surpass $24 million this year.

But to keep this growth in perspective, Charlie McEhlone says Australia is at a large disadvantage compared to New Zealand, which enjoys a free trade agreement.

...

'Whole milk powder, the major imported product of dairy, they're controlling about 98 per cent of that market, so they're controlling with that tariff advantage.'

...

Source: Sarina Locke, ABC Rural, 4 June 2013

2 Read the article and complete the following questions.

a Explain the need for milk powder in China.

____

____

b Outline how an Australian town is able to meet that need.

____

____

c Based on the article, discuss what steps could be taken to make trade with China easier.

____

____

12.4 Definitions: Changing technology

Knowledge and understanding

 **verbal–linguistic**

1 Circle the correct term in the sentences below.

a Core / Periphery: the countries and regions of the world with the greatest wealth and power

b Globalisation / e-commerce: the buying and selling of products or services using the internet

c Virtual space / Globalisation: a computer environment that can simulate a physical presence in places in the real world or imagined worlds

d Globals / Mobals: people who have a global outlook; they have influence because of the nature of the jobs they do, and have enough wealth to be able to travel freely

e Interconnectedness / Globalisation: being interconnected or linked to other people or places culturally, economically or socially

2 Draw a line to connect these terms with their definition.

Term	Definition
social media	area surrounding a place that is linked to that place through lines of exchange or interaction.
virtual space	people who move from their original place of living to somewhere else, solely for a better job or a better life for their family
virtual community	the countries and regions of the world outside the core
periphery	communication technologies through which users create online communities to share information, ideas, personal messages, and other content forms of content (e.g. videos)
mobals	a community of people sharing common interests, ideas and feelings over the internet or using other collaborative technologies
hinterland	a computer environment that can simulate a physical presence in places in the real world or imagined worlds

3 Use the following words in a short paragraph to show their meaning.

a local: ______________________________

____

b globals: ______________________________

____

c mobals: ______________________________

____

13.1 Production and consumption of goods

Knowledge and understanding • Geographical skills

verbal–linguistic • visual–spatial • logical–mathematical

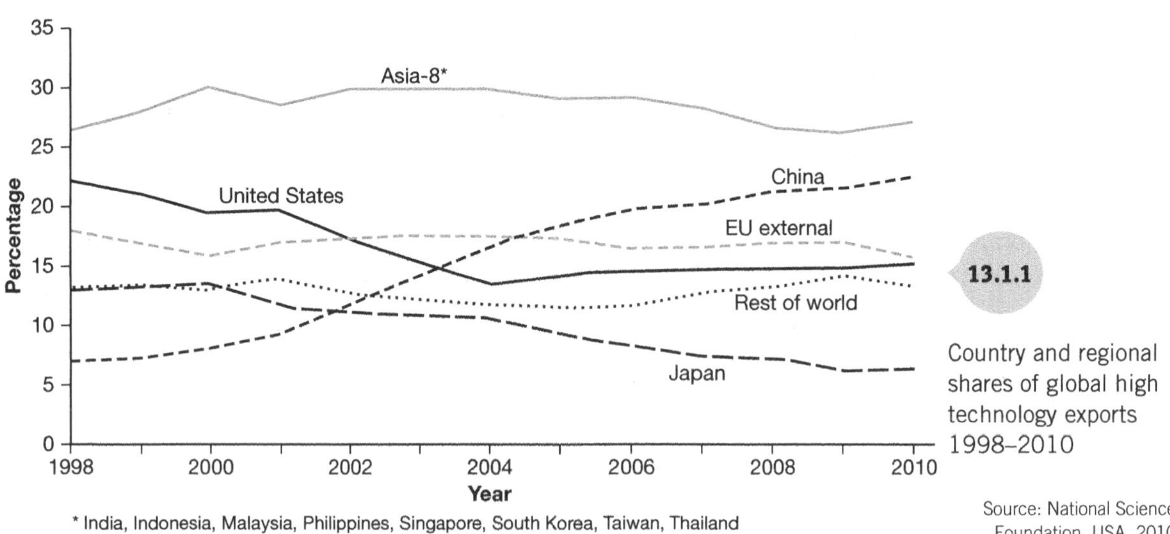

13.1.1

Country and regional shares of global high technology exports 1998–2010

Source: National Science Foundation, USA, 2010

Use Figure 13.1.1 to complete the following questions.

1 Which region had the greatest share of global high technology exports in 2010?

____

2 Which regions or countries have decreased their share of global high technology exports between 1998 and 2010?

____

3 Describe the changes in importance of China as an exporter of high technology parts.

____

____

____

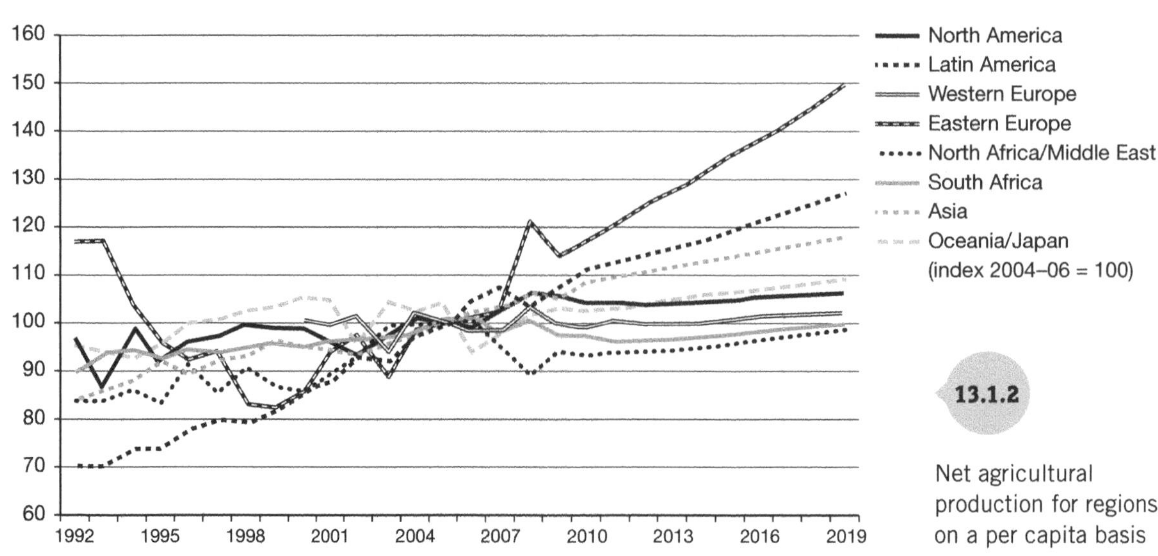

13.1.2

Net agricultural production for regions on a per capita basis

Source: OECD–FAO Agricultural Outlook

Use Figure 13.1.2 to complete the following questions.

4 Rank the regions according to the largest net production predicted in 2019.

____

____

5 Compare the agricultural production of Eastern and Western Europe from 1992 to 2019.

____

____

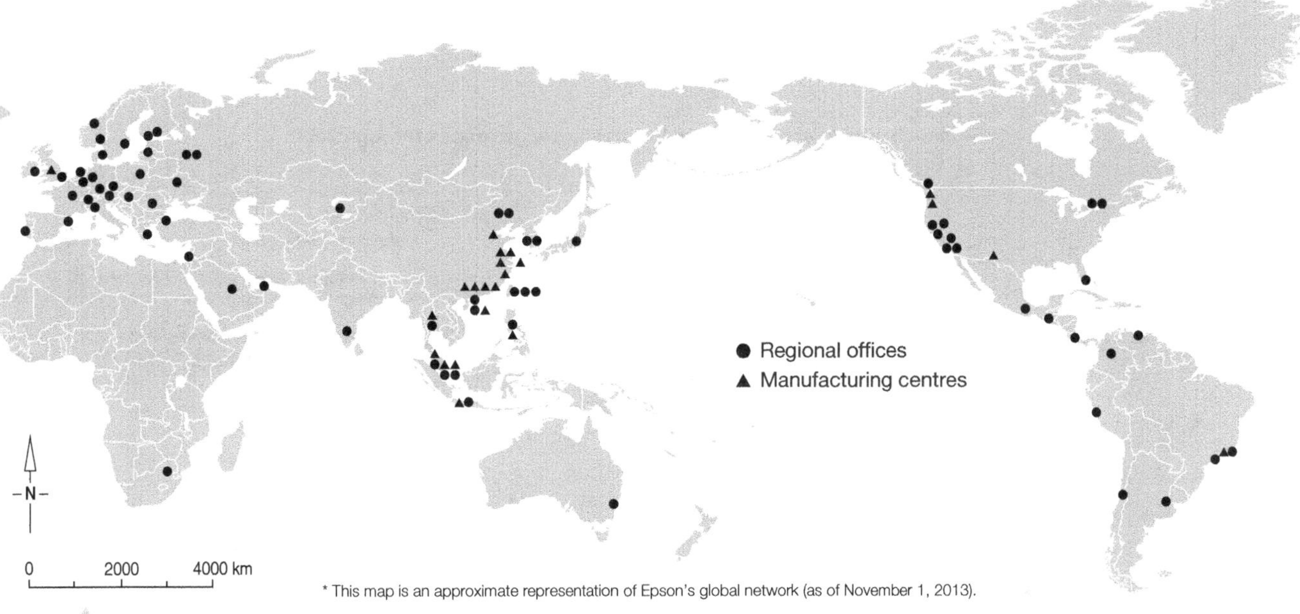

Source: Epson

13.1.3 Global locations of Epson offices and facilities

Epson is a Japanese multinational company that specialises in the production of items such as printers, digital projectors, robotics and medical equipment. Use Figure 13.1.3 to help complete the following questions.

6 Suggest why Epson has regional offices in North America, Europe and Asia.

____

____

7 List the advantages and disadvantages of having manufacturing centres in various locations throughout the world.

____

____

____

____

There are Epson sales offices in more countries than countries with manufacturing centres. For example, São Paulo, Brazil, is the location for the only Epson manufacturing centre in Central and South America. The regional office at Long Beach, USA, would also have control over production and sales in Central and South America.

8 Explain how an increase and decrease in sales of Epson products throughout Central and South America might affect other places.

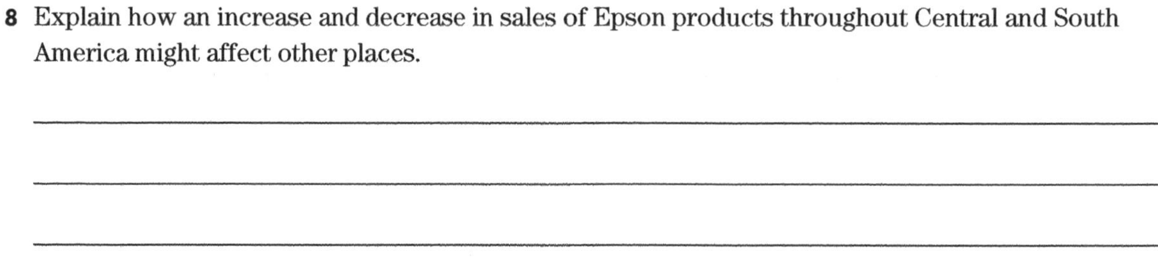

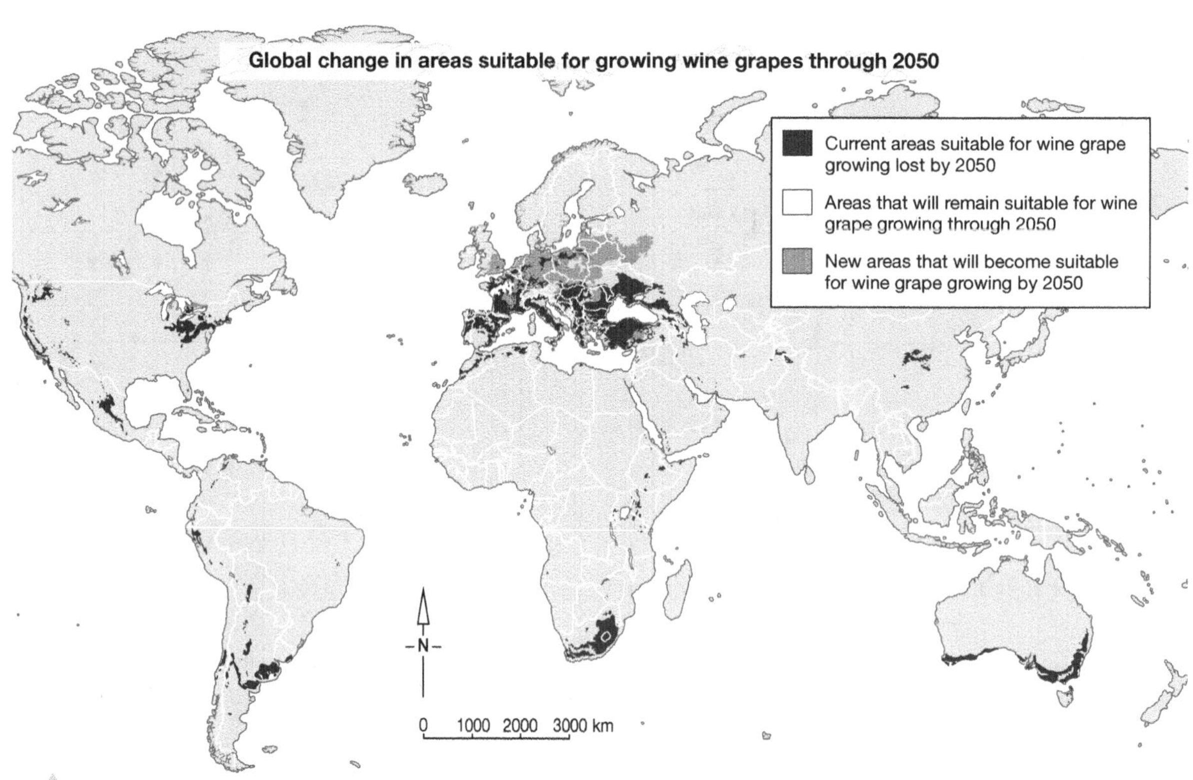

Source: Conservation International, 2013

13.1.4 Map showing current and projected areas for wine growing suitability

The production of goods may be affected by changes to the environment, such as through climate change. Study the predicted changes by 2050 on the areas suitable for growing wine grapes.

9 Describe how wine growing areas in Europe may be affected by climate change.

10 Discuss the significance of climate change on wine growing areas in Australia.

13.2 Food products and biodiversity

Knowledge and understanding • Geographical skills

mi visual–spatial • verbal–linguistic

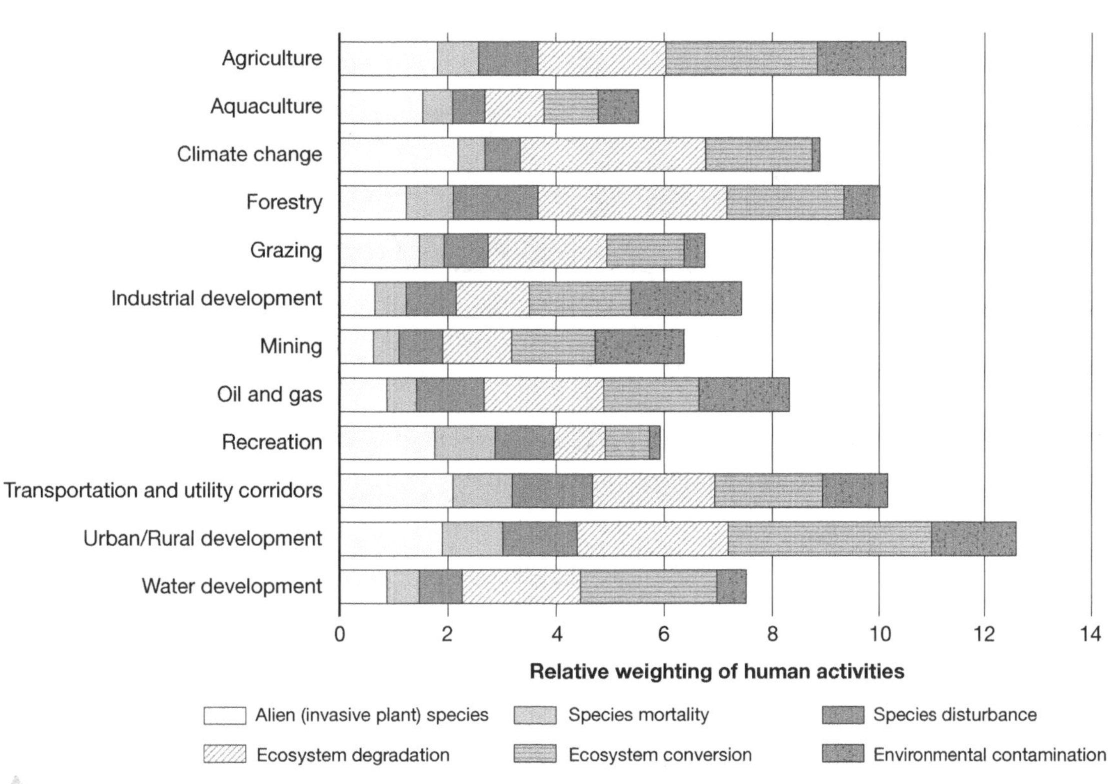

13.2.1 The environmental impact of human activities on plants

Use Figure 13.2.1 to compare the environmental impact of agriculture with other human impacts.

1 Compare the impact of agriculture on alien species with that of mining.

___

2 Compare the impact of agriculture on ecosystem degradation with that of forestry.

___

3 Compare the impact of agriculture on ecosystem contamination with that of grazing and aquaculture.

___

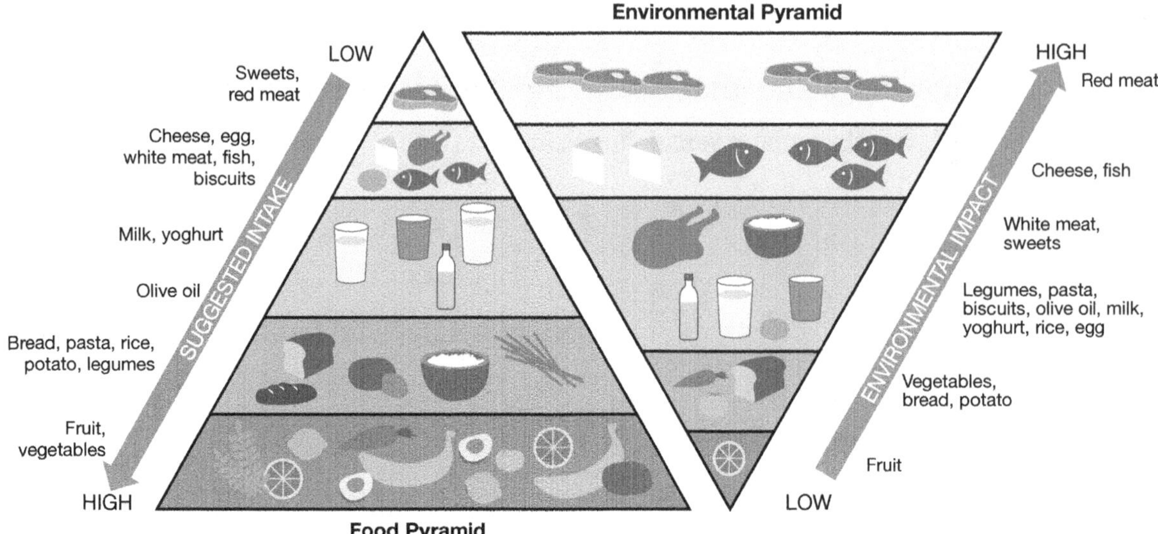

13.2.2 Diagram showing the environmental pyramid compared to the food pyramid

4 A student sees the diagram in Figure 13.2.2 and concludes that the foods we should have the least of are the ones that have the greatest impact on the environment. Explain why you think the student is correct or incorrect.

____

____

____

____

____

5 Suggest a daily menu that is healthy for you and has a low impact on the environment.

a breakfast ____

____

____

b lunch ____

____

____

c dinner ____

____

____

d snacks ____

____

____

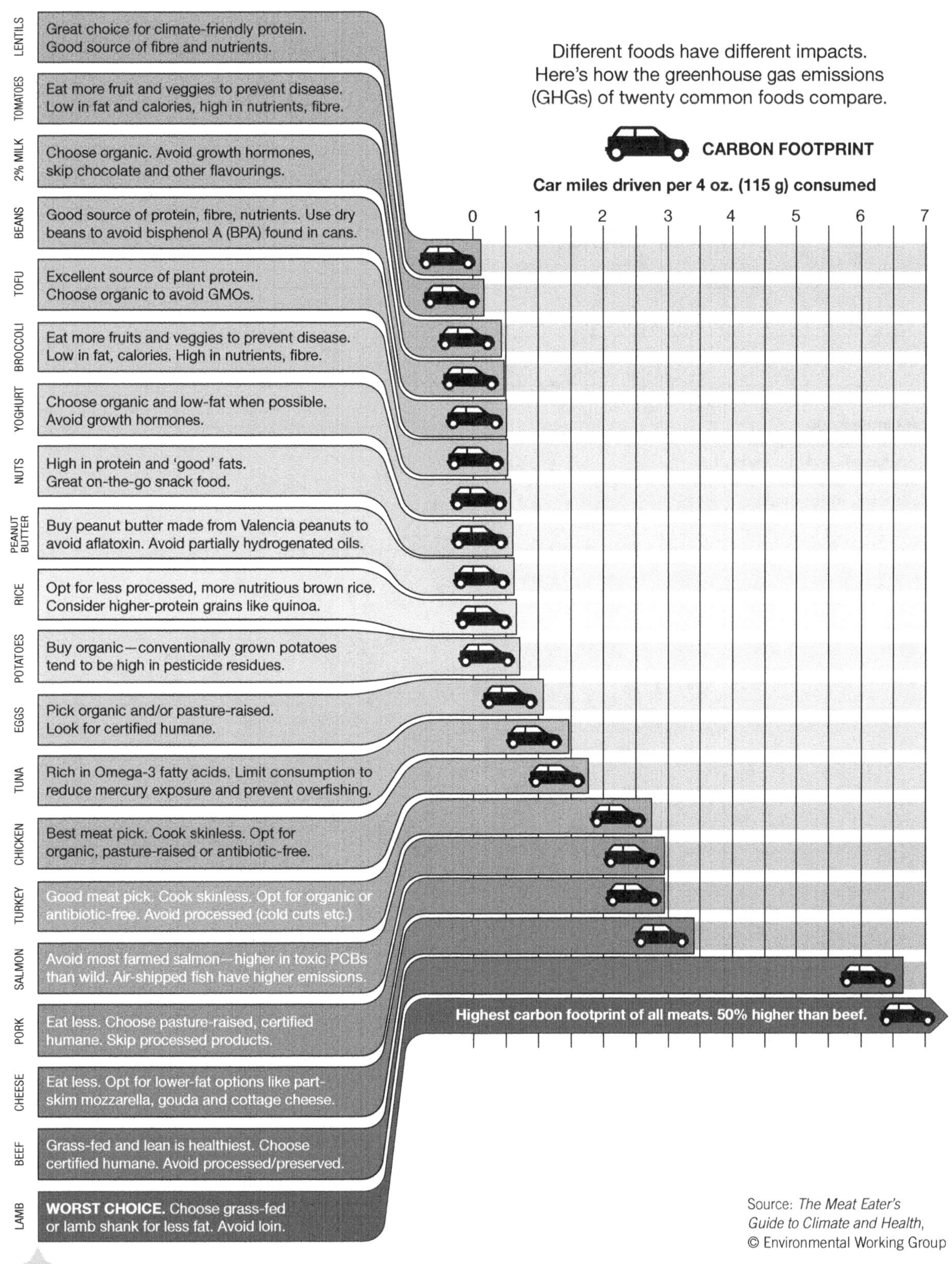

13.2.3 The carbon footprint of different food groups

6 List the foods you ate for your most recent dinner.

____

____

Refer to Figure 13.2.3 to help answer the following questions.

7 Ignoring the weight of the food, describe the carbon footprint of your dinner.

____

____

____

8 Suggest a menu that would be have a lower carbon footprint than your last dinner.

____

____

____

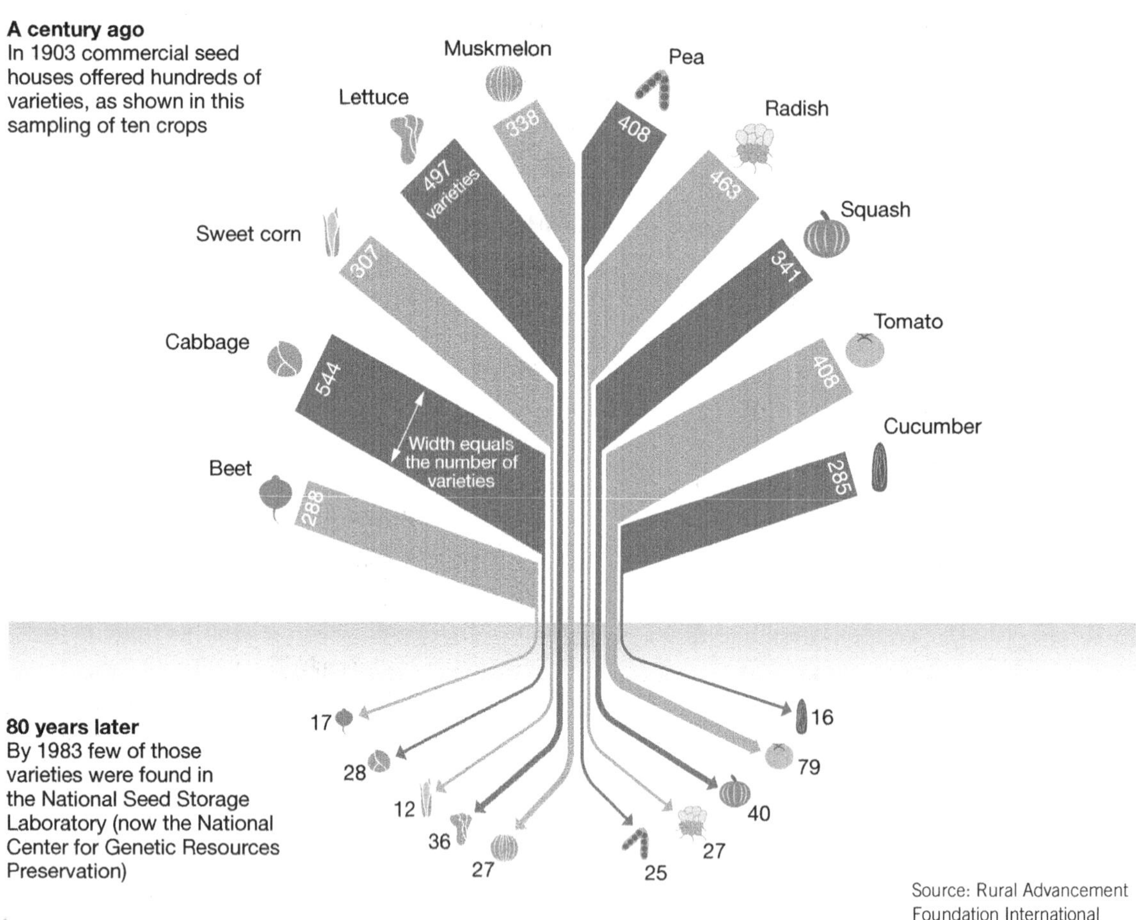

13.2.4 The varieties of food crops in 1903 compared to 1983

Refer to Figure 13.2.4 to answer the following questions.

9 Explain the environmental impact of fewer varieties of crops.

____

____

10 State one disadvantage of having fewer species of crops available.

____

____